Star of the Four Kingdoms

I would like to show my depth of gratitude to the following people and organizations for their help and contributions.

NLB and The Commissioners of Northern Lighthouses
www.nlb.org.uk
Photographs of the HQ, Ruvaal, Holburn Head, Inchkeith Unlit, Rhinns of Islay Lighthouse, The Flannan Isles & the Lighthouse Logo,
Virginia Mayes-Wright, Director of The Lighthouse Museum Kinnaird Head.
www.lighthousemuseum.org.uk
Photographs; Inchkeith Lit, Ailsa Craig, Killantringan, Rhinns of Islay from Portnahaven and Sule Skerry Lighthouses
Bob Jones for the Photograph of Barra Head lighthouse
Vanessa Langley for the Photographs of Ailsa Craig today, and the foghorn.
Special thanks goes to former Principal Keeper
Photographs of the Rhinns of Islay, Model of a launch and the Works Order Documentation.
Cover & Mull Photographs courtesy of
Steve Hardy
South Rhinns Community Development Trust
www.mull-of-galloway.co.uk

COMMISSIONERS · OF · NORTHERN · LIGHTHOUSES
IN SALUTEM OMNIUM
·1786·

This book is dedicated to my wife Margaret
and
our children Karen, Kirsty and Gavin

Best Wishes

Peter J. Hill

ISBN: 978-1-84914-012-8

Preface

Imagine if you will, a young man called Peter Hill born on the fifth of June 1953, brought up in Glasgow, later to study Art at University. At 19 he is somewhat at a loss; when in 1973 his studies come to a premature end. He joins the lighthouse service as a relieving keeper, a temporary position to kick start a pathway to an undecided future.

Now take another Peter Hill born 12 days earlier on the 24th May 1953, 3000 miles away in Kenya at the time of the Mau Mau uprising. Not related in any way to the former nor destined to meet him but yet uncannily their paths were to follow however briefly a similar course. He too at a loss joins the same lighthouse service six years later not as a relieving keeper but as a permanent full time keeper.

Peter Hill from Glasgow is compelled to write of his exploits and the people he met during his brief time as a lighthouse keeper in the book Stargazing; but what if he had chosen to take a different path and elected to remain in the service as an Assistant Lighthouse Keeper would not the tapestry and colour of his world makes for perhaps a more interesting read.

Peter Hill from Kenya chose the second path and with the same compelling force but with perhaps slightly different motivations has chosen to write his memoirs of his time as a Lighthouse Keeper. So there must be Lighthouse Keepers past and present who would have their own stories to tell. If they were put together they would chronicle our way of life. Perhaps there may come a time when someone with the insight and the inclination will gather those memories and make a book far more worthy of the story.

There are still Lighthouses and Keepers who attend them but this book is dedicated to the time and to the Keepers just before automation cast them away to history books.

I am often asked," What made you think of becoming a Lighthouse Keeper?" This question, I think is asked because of a notion that the life of a Keeper is somewhat romantic, dramatic or as surreally characterised in novels.

I suppose the answer could be that it's a bit of all those things and a lot more besides. Why and what makes a doctor become a doctor, or an airline pilot an airline pilot?

We are all driven on one way or another, whether we are shaped and influenced by our upbringing or whether by circumstance, motivation guides us into selecting our paths in life.

This story will help answer those questions and show you what life was like for me as a Lightkeeper. It's not just about me though; you will be able to see that we were a community bonded together by the nature of the service that we provided, and far from being extraordinary people, our lives followed the same patterns as everyone else. Perhaps we did have more than our fair share of comedy, drama and perhaps we did experience events that sound surreal, but to us they were all possible given the very nature of the service. In the 13 years of my service I have found myself in some amusing interesting and unusual circumstances which surely must have been experienced by the generations of Keepers that preceded me, all bear testament to the uniqueness of the service which sadly for both Peter Hill's, our families and those Keepers still alive, are but cherished memories.

This then is my story.

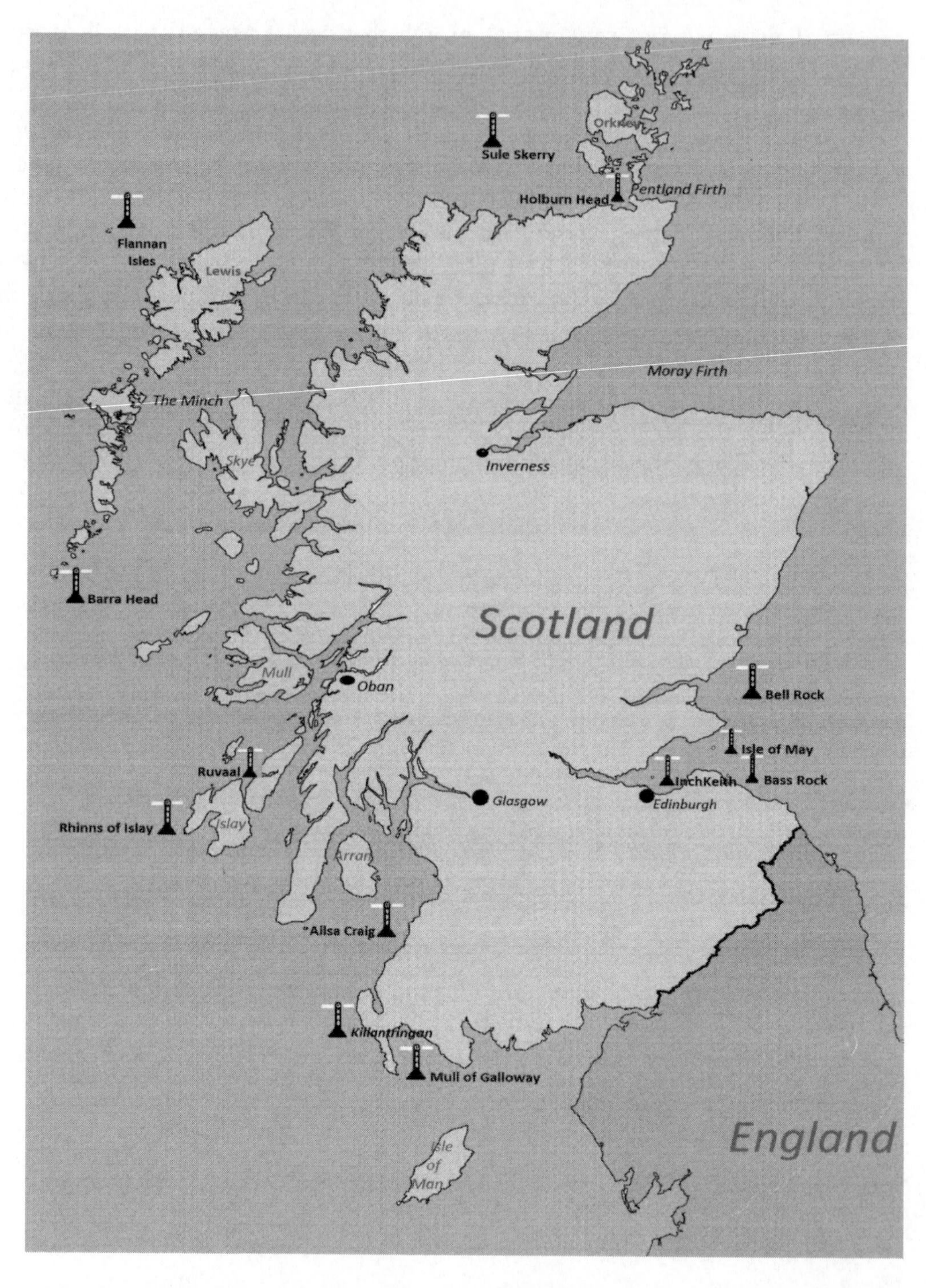

This map shows the general location of the lighthouses I served on and includes the Flannan Isles and the Rock Stations of the Firth of Forth

Chapters

I.	The Spare Man
II.	Killantringan Lighthouse
III.	At a Loss
IV.	Sule Skerry
V.	Ruvaal
VI.	Barra Head
VII.	Holburn Head
VIII.	While you were away (Wait till your father gets home!)
IX.	The Move to Edinburgh & The Relief
X.	InchKeith. The Island & The Station
XI.	Winter & Summer Work & The Commissioners Visit
XII.	The Wreck of The Swither & The Helicopter Relief
XIII.	The Riddle of The Bones & The Winter of Discontent
XIV.	While you were away at InchKeith
XV.	The Move to The Mull of Galloway & The Station
XVI.	Fixing the Problems
XVII.	New Arrivals at the Station
XVIII.	Sprucing up the Station
XIX.	Striking Oil & The New Central Heating
XX.	High Days & Holidays
XXI.	CB Radio & The Russian Ships
XXII.	A True Family Station & The Price of Progress
XXIII.	Ailsa Craig. The Island & The Station
XXIV.	All Ashore Who Are Going Ashore
XXV.	Adventures With Donald
XXVI.	Changes at the Rock
XXVII.	While you were away at the Craig
XXVIII.	Islay & The Rhinns
XXIX.	In Sickness & In Health
XXX.	Last Days in the Service

84 George Street Edinburgh
HQ of the Commissioners for Northern Lighthouses

Logo and HQ pictures courtesy of NLB.org.uk

Lighthouse above the very door

Chapter I
The Spare Man

To this day the adverts the Northern Lighthouse Board place on recruitment pages of newspapers show the image of a Lighthouse; it was this image that caught my attention and aroused my imagination. Reading on to the rewards, I began to realise that this position could provide both a financial and a somewhat stable future for us; as well providing a good salary, housing would be provided rent free, rate free, and fuel cost free. It would mean a wrench from Margaret's family again and we would not be able to put down roots having to move every few years or so. Therefore I was uncertain if Margaret would even consider the idea, so it was with trepidation that I showed her the advert.

Pondering awhile and after what seemed an age she said, "Go for it." I knew there would be a lot more to it than just that, but it was what I considered was a step in the right direction. Whenever there were important decisions to be made, we drew up a list of the pro's and con's, here the pro's far outweighed the con's; I was definitely biased in favour but whatever deep down reason Margaret had for agreeing, she kept it to herself. In the end, I suppose it must always come down to one person to make the final decision, so I wrote for an application.

As with most employment applications, it was never certain that it would be accepted let alone get to the next stage. However, I had hoped that my service with the Army would stand me in good stead. I should have had reservations about that though because I did apply once to join the Ayrshire Constabulary just before I left the Army.
It even got as far the interview stage, but my hopes were dashed when my height was re-measured revealing me to be only five feet seven and three quarters; a quarter inch shy of the required height. My disappointment at being rejected caused me to voice my objection. "How on earth can you justify that decision, when I have faced far greater risk policing the streets of Northern Ireland?" The decision remained unchanged, so maybe that was all part of the test, just to gauge how I would react and I failed because

of my remarks. Well I'll never know but I have no regrets and still have the tendency to speak my mind.

It seemed as though my application had reached an impasse; even after the glowing references given by a Doctor and Justice of the Peace. So I wrote a pleading letter for the Board to consider my application. It could have been the letter or natural timing that produced the desired results and a letter requesting me to attend an interview at 84 George Street Edinburgh; the Headquarters of the Board duly arrived.

I had been to Edinburgh only once before with Margaret; when we were courting, then, we took in the usual tourist sights; Princes Street, the Castle, The Royal Mile, Holyrood etc. I did however get the feel of the layout of Edinburgh, and I had a vague idea where George Street was, but just to make sure, the Board kindly supplied me with a hand drawn map. It was basic enough to show the directions from both bus and train stations; I elected to go by train.

I like to arrive at an appointment before time, so I set off early and chose to go the Glasgow Central to Edinburgh Waverley route, giving me plenty of time to compose my thoughts for the interview.

My experience of interviews was limited. I left school at 15 with no qualifications. I had only three jobs before joining the Army. My first was on a poultry farm earning £5 per week. I think my employer was only too happy to find anyone to fill the position; and being that age, it was a relatively safe bet that I would not be of unsavoury character. My second was on a mixed arable dairy farm, at an increase of £1 per week. I enjoyed the work but knew that there was no real prospect of ever progressing to anything more than just labourer; both jobs totalled less than a year...

My mother died when I was ten and my father, a man in his fifties, found himself with five children to bring up. My eldest brother Bryan was at the stage of virtual independence and I seem to remember seeing very little of him. How much he contributed to the family income I do not know. I do know that at the age of 13, I made the conscious decision, however naive, to leave school as soon as the law would allow and find work to bring in an income to the family. It was not as if I felt that I was sacrificing an education, for I was not academically inclined and would not have progressed on to further education. With hindsight this was the single most regrettable decision I have ever made however, if, as is my belief, I had chosen any other path, it would not be these memoirs I would be writing, so I am at peace with that decision.

With another elder brother Allan and my sister Megan both contributing to the family, my father did not put pressure on me to increase mine. At this time I did not feel particularly ambitious and was uncertain what I wanted out of life, so my future seemed to lack direction

Allan helped me get a job in the factory where he worked, doubling my wage in the process. Ambling along in this job, I became proficient at making number plates, repairing car radiators and fuel tanks and all other new tasks that the company would diverse in, but there was an underlying feeling that there must be more to life than this. Perhaps everyone goes through this phase in his or her life. I had always thought that to have an ambition you must have a definite goal, a point at which to aim for. This feeling for me was more of emptiness, something was missing, I did not know what, but I yearned to be fulfilled.

Until I knew what I wanted to do with my life, I felt the best thing for me to do was to join the Army. After all the Army had provided my father with a way of life for nearly 30 years, and it was instrumental in my very existence. Pop, as my father preferred being called did not actively encourage me to join the Army, but then again he spoke little about his life let alone his Army exploits.

We knew little about his past. He'd said he was born in Reading in 1906. We recently unearthed that he was actually born in Thatcham, outside Reading. He was the eldest of three brothers and had a strict Anglican upbringing. He bore the brunt of disciplinary punishment, which he resented greatly. At the age of 14 he ran away to become a boy soldier with the Coldstream Guards, he was made redundant in the great demobilisation of 1927, called up again in 1940 he served with the RAOC during the war and rose to rank of WO II until his retirement in 1957.

When asked, "What did you do on the war dad?" his only reply was, "I don't know I was too busy running." We did get one brief story though; about a time when he got lost and wondered perilously close to enemy lines. Ahead of him in the gloom of night lay a bridge crossing a wadi that he thought he recognised would take him back to the safety of the lines. Half way across he saw a figure approaching, in an instant he recognised the uniform of the German Afrika Korps and beat a hasty retreat in the opposite direction just as fast as his legs would carry him.

As no shots followed him, the German had the same idea. "And that, children, was the closest I ever came to a German," he said. I don't believe for a moment that was the total sum of his experiences during the war, but it was his nature to keep things to himself.

He had been married before; his 1st wife bore him five children, three girls and two boys. He did say once, "If I could get you all together, we could form a football team." It sadly never happened, for we have never met our half brothers and sisters. It must have pained my father deeply having to split up his first family. The details are vague; but apparently while he was serving away from home, the children had been neglected. He returned to find them in such a state; it forced him into taking the drastic steps of putting the three eldest girls into care and the boys out for adoption. This could not have taken place without due process so the circumstances must have been exceptional. He only ever made one reference to their mother and that was to say she wore Wellington boots in bed during a thunderstorm to ward off lightning bolts

Perhaps the event was so painful that it had a great influence on him later when our mother died. I don't believe he could have lived long seeing the breakup of his family for a second time. So at the age of 57, when a lot of men would have their thoughts drawn towards retirement, his thoughts were to bringing up the five of us single-handed.

I may have inherited some of his characteristics but it was neither that nor his influence that made me look toward the Army to provide a career. What drove me most is that I began to see the possibilities that the Army might offer, especially when the local newspaper had an advertisement for a locally based Artillery Regiment looking for recruits with the potential to become radar operators. As things turned out the Regiment was permanently based in Germany with the occasional sojourns to Northern Ireland and the annual trip to firing camp. What is more I never became a radar operator.

My thoughts returned to the impending interview. What kind of questions will they ask? I began to reassure myself with confidence and self-esteem that I would present myself as the candidate they were looking for.

I had not been paying too much attention to the passing scenery until it suddenly dawned on me that it was no longer passing. We had not stopped at a station! After enquiry from a fellow passenger I discovered we had been stopped for nearly ten minutes. Although I'm not one for panicking, the thought of being late for the interview and the consequences were giving rise for concern. There was a gap of 40 minutes between the scheduled arrival at Waverley Station and my appointment. Ample time I thought, given that even at a slow march it would only take 15 minutes to get to there. What was it Burns said about ..."The best laid plans of mice and men?" With the unscheduled stop both times were set on collision course, I knew now that it was likely to be a sprint finish to get to 84 George Street on time. And it was breathless that I arrived on the doorstep of the Board. I had hardly time to notice the grandeur of the facade, broken only by the black veneer of the large panelled doors with their gleaming brass handles and the equally gleaming nameplate; Commissioners for Northern Lighthouses, let alone the stone modelling of the same lighthouse depicted in the advert, above the very door it had brought me to.

Polished brass and woodwork were evident throughout the foyer, equally as smart was the Commissionaire who politely asked my business and ushered me to a plush leather chair to await someone from Personnel who would take me upstairs.

Not meaning to patronise, but more to give historical account, the difference between Commissionaire and Commissioner were about £40,000 much to the dismay of the gentleman who has just ushered me to my seat ...

The Northern Lighthouse Board (Commissioners for Northern Lighthouses) was established in 1786 by an Act of Parliament in order to oversee and manage the proposed future development and establishment of navigational aids around the coast of Scotland (later to include the Isle of Man). The men Commissioned by Crown and Parliament were usually taken from the Sheriff's Principal of the coastal shires of Scotland and other dignitary's such as Provost hence the differential in salary between the two.

Having the chance to catch my breath and look around, I could not help but notice the spit and polish of my surroundings. With my military experience, it was something I was accustomed to, but here you could feel the difference; in the Army one tends to carry out Bullshit (if you'll pardon the euphemism) with disdain, but in this building there seemed to be this overwhelming sense of pride that effused from every polished surface. Was this a foretaste of things to come?

I am taken up to the second floor, expecting to see similar decor to the foyer, I was surprised to see the building had recently undergone modernisation and the transformation par with the style of the late twentieth century.

The interview went well and I felt as though I had given good account of my motivations and abilities. At the end of the interview I was asked to wait in the Staff canteen for the decision to be made on my suitability to proceed to the next stage.

I was more nervous in that canteen than at any other time, given the time of day and the lack of a more suitable sustenance I had elected to eat a jam doughnut with my cup of tea; bad choice ... You've guessed it, I had no sooner picked up the said doughnut and taken a bite, when the over laden doughnut struck back, covering not only my chin, but my lap in the sticky substance. The chin was easy to deal with, with a wipe from the paper napkin. My trousers would not be so easy to deal with without leaving any telltale marks.

Just as I had begun to roughly wipe off the jam with a napkin, I was called back to the interview room. My composure all but evaporated in that instance. Making the excuse of needing the toilet, which out of necessity was not a complete fabrication, I endeavoured to make good with the help of a towel and some water; second bad choice .... It now looked as though I had been overcome by the moment, coupled with my earlier excuse the evidence seemed all too convincing. I doubted if anyone would make enquiries and hoped that it would not be too noticeable.

It became obvious that the dampened patch around my crotch area was drawing some attention. Grasping the bull by the horns and knowing I could be damned either way, I decided to come clean (sorry for the unintentional pun). Whatever the result, it was not held against me, for I had been provisionally accepted into the service.

The acceptance was subject to my passing a Medical, so after bidding my farewells and receiving yet another set of directions that would take me to the Board's Doctor, about a half mile away; I passed through the front doors turning left to walk to the Surgery of Dr. Sir John Halliday-Croombe.

It did not take much deduction to work out that I would soon be entering the world of private medical practice, a first time experience for me and if the experience of what followed is typical, I'm glad that it was my last.

Moray Place like many other streets in the area was built in the Georgian era to show off the growing affluence of the Capital city, that affluence still evident and not lost to the passage of time. The grand sweep of the Georgian Circle every bit as resplendent as the one in Bath, with each house identical in form so as not to lose the magnificence of the whole, the only mark of the individuality of the occupants was the colour of the doors.

A brass nameplate assured me I was at the correct address and I attempted to enter, the handle did not move and the door was firmly shut against me. Looking for a bell or some other means of attracting attention to my presence, I noticed a chrome plate set in the wall of the portico, as it had a button, and punched holes like the mouth piece of a telephone I deduced that I should press this button and gain entry if I satisfied the respondent to my identity. I felt a bit like Alice in Wonderland only without the benefit of instructions. Such entry systems are commonplace now but this was my first encounter and one I never expected to see at a Doctors. After dutifully pressing the button, a voice from the mouthpiece requested my name and within a second a buzz and click allowed me to push the door and gain entry.

The interior of the house lacked the residential trappings but nevertheless the impression was that this had not always been a surgery and only minor refurbishment would return it to its original status.

The receptionist pointed to a large Chesterfield where it was thought I might lose myself, it only added to my growing discomfort. There is what some describe as a clinical coldness, caused by a mixture of apprehension and surroundings, not that there was anything physically cold about my present surroundings, it was more of an atmosphere. I felt myself in a place where I was not wanted. I was upstairs when I should have been downstairs or maybe it was just that I was not personally paying the bill, whatever the reason, I was still ill at ease, and this disease was certainly not going to be cured by the eminent Doctor who was about to examine me.

I was about to undergo physical examination and verbal cross examination, that even De' Torquemada of the Spanish Inquisition would have been proud. This of course was gross exaggeration but his lack of bedside manner was not. Once he had finished poking, prodding and compiling a medical history of every complaint since childhood, which I might add, I was made to feel personally responsible for; I was more than happy to get dressed to put an end to the ordeal.

Little was said to show that some medical impediment might spoil my chances of becoming a lighthouse keeper, so I correctly assumed that this was another stage passed on my quest.

Waiting for important news or forthcoming events without showing signs of impatience, has never been one of my strong points. Once the wheels are set in motion, I like to move on. It was the same when I left the Army...

The record shows that I left the Army under the conditions of discharge by purchase, meaning I bought myself out of the contract, of course it does not say why. I will only say that my decision to leave the Army was because; it could no longer offer me the chance of advancement and direction that I had found or would it allow me to fulfil my impending marital obligations. I'd met Margaret while I was serving in the Army and shortly after returning from a four month tour of Northern Ireland. The monotony of camp life in Germany was getting to most of us, some lads in one of the other Troops had, for a laugh, submitted a guy's name to a girls' magazine for a pen pal. He had over 300 responses; there was no way he could answer them all, so he gave me a half dozen to select a pen pal from.

When I opened Margaret's letter a small photograph fell to the floor, it appears from the moment I saw her I was smitten. Within three months of writing to her, I proposed, not ever having seen her in the flesh so to speak. So there I was having requested and completed the discharge papers and paid the required fee, I had to wait for the MOD's pleasure, I'd even received the returned as a paid cheque, and an order to return to Depot. I could not officially leave the Regiment until an official discharge notice was received at Regt HQ. It is customary to always obey the last order and as I was not ordered not to leave the regiment, I decide to return to Depot, which for us in the Artillery is based at Woolwich.

Technically I was AWOL but as I had reported to Depot on time it was questionable whether any infringement of Military Law had occurred. The Army did however show its displeasure by retaining my services for a few weeks longer than I think was necessary, it could have been a lot worse, I could have been sent back to Germany to face charges of being absent without leave and the possibility of even greater delays while the matter was sorted out...

Bearing all this in mind I felt that the virtue of patience should be observed on this occasion and wait quietly for word from the Board.

It came as a letter requesting my attendance at 84 where I would be told which Lighthouse to go to. Accompanying the letter was a kind of travel warrant; similar to those issued by the military, which could be exchanged for a valid ticket, not having to pay for the ticket this time was my first feeling of acceptance into the service. As I would be going directly from Edinburgh to my first assignment, I packed enough clothes for what currently was an indeterminate period away from home.

Kissing and holding Margaret and the girls close in a way that I thought would be most reassuring, I bade farewell, comforted by a backward glance seeing that Margaret was dry eyed, where previously even the thought of a brief separation had brought tears. I believe Margaret had resigned herself to getting used to the idea of separation.

The one great stumbling block to progressing on to becoming a Lighthouse Keeper is to be able to accept and live with separation from ones family and already Margaret had shown the first signs of this acceptance. I was so overwhelmingly proud of her at that moment; it was I that had to wipe the moisture from my eyes.

Not risking another train delay, I chose to take the more conventional route via Glasgow Queen Street and duly arrived five minutes before my scheduled appointment.

I was due to be welcomed into the service by the General Manager; Commander McKay, but he would be delayed for an hour, so I was sent to be measured up for my uniform. Following the lines of expectation by my previous dealings with the Board, I found my way to the Board's tailor not far down the Street. In the window the manikins were dressed in suits of the finest quality, standing out from them all was one dressed in naval uniform. I need not fear disappointment, for I was to learn that my own uniform would be made to the same exacting standard. I had been measured for a uniform before, but never with the care that was shown on this day.

Entering the General Managers office was like a trip into the past, none of the modern refurbishment here, the history of the service evident everywhere. On the walls portraits of people yet unknown and pictures of ships of steam and sail whose names were barely distinguishable, only one ... The Hesperus ... recognised from distant memory. Around the small tables, windowsills and on the large desk before me, were items of Lighthouse paraphernalia.

Commander McKay looked well suited to his surroundings and I could picture him giving the same speech in 1879, a hundred years before. His voice though soft spoken had the authoritative quality to it that came from years of commanding others. I was at ease.

Hearing a history of the service, it brought home to me the importance of the role that I was undertaking, but never did I think for one moment that I would not be able to meet those expectations. His parting words, with the shake of my hand; "Welcome Peter to the Lighthouse Service, you are now Supernumerary Lightkeeper, I hope that we meet again soon, when you are appointed Assistant."

Returning to the main office, I was given a pack; which included details of what lighthouse to go to and my contact there, another travel warrant, my service regulation book, a booklet on health and safety and a letter detailing the terms and conditions of the probationary period under the title Supernumerary Lightkeeper. I had also to sign some documentation and fill in next of kin details and for a second time in my life, a copy of the official secrets act.

Aboard the train on the return journey, for indeed it was a return journey, retracing my exact steps of the morning back to Ayr, where my journey would extend on to Stranraer; Killantringan Lighthouse was to be my final destination.

Later I looked up the meaning of the word ... Supernumerary ... roughly translated ... Spare or extra.

I had ample time on the journey to read through the Service Regulations in an attempt to glean an insight to what lay ahead for me, however, with terminology as yet still beyond my ken, I resolved myself, that all would be revealed in time...

From time to time through the story I will give snippets from the regulation book and put them into context of how they affected me as Lightkeeper and the family.

Just to give credence to the uniqueness of the service, the word Lightkeeper does not appear in most dictionaries. It is or was in general use to describe our status within the service and our formal title in the service regulations. Whether our rank was Principal, Assistant or Supernumerary our address for correspondence to Board would be Lightkeeper and from henceforth you will see this address...

Page one of the regulations:

SERVICE REGULATIONS
LIGHTHOUSES

These Regulations, having been approved by the Commissioners, are issued for the information and guidance of all concerned.

Throughout these Regulations the expression "Principal Lightkeeper" is also to be taken to mean the Senior Assistant Lightkeeper in charge of the Station in the absence of the Principal.

... Then the signature...

W. Alistair Robertson
General Manager

1st December 1965

... Obviously there must have been regulations in existence before this date and they would have been amended on numerous occasions, but apart from a few amendments in this book and a change in the General Manager these were the regulations in force until my departure from the service...

SECTION A. GENERAL

A.1. PRINCIPAL LIGHTKEEPERS. - A Principal Lightkeeper is placed in charge of each Station and is responsible for all the property of the Commissioners at the Station. He is also responsible for conducting official correspondence in connection with the Station, for the cleanliness and good order of the Station and for the good order of the Station generally and for the due observance of the Regulations appertaining to Lightkeepers and of such Orders and Instructions as may be issued from time to time. Any case of disobedience or disrespect on the part Assistants should at once be reported to the General Manager.

The Principal Lightkeeper may at any reasonable time enter and inspect an Assistant's dwelling if he has any doubt regarding its cleanliness and good order. Where there is an empty house he must ensure that all rooms are properly ventilated, heated and kept clean and orderly.

... In short the Buck Stops with Him.

At no time throughout my service did the Principal ever conduct a spot inspection of our house, nor did I ever hear of an occasion other than hearsay of a Principal having to do so. The very thought of having what amounts to a stranger come to inspect your house would I expect be like a Red Rag to a Bull to most women, and initially Margaret's reaction was the same, but on reflection, she deemed that condition of the house at anytime was nothing to feel shameful about ...

A.2 (a). ASSISTANT LIGHTKEEPERS. - Assistant Lightkeepers are subject to the control to the control of the Principal Lightkeeper, and must obey his orders. If an Assistant Lightkeeper feels aggrieved by any order of his Principal, he may appeal through the General Manager to the Commissioners, but he must obey the order meantime.
The Senior Assistant will assume charge of the Station when the Principal Lightkeeper is on leave or respite from duty. But in the event of an emergency or breach of discipline when the Principal Keeper is at the Station the matter should be referred to him in the first instance

A.2 (b). LOCAL ASSISTANT LIGHTKEEPERS. - Local Assistant Lightkeepers are subject to the same discipline and are required to undertake the same duties as Assistant Lightkeepers. They are liable for duty at any of the Commissioners' lighthouses in their locality but are normally appointed permanently to a particular lighthouse after the satisfactory probationary period of one year.

A.3. SUPERNUMERARY LIGHTKEEPERS. - A new entrant to the Service is appointed a Supernumerary Lightkeeper, and is sent to one or more Lighthouses for a probationary training period; he may also be required to attend a training course.
On completion of this training period to the satisfaction of the Commissioners he will be made Assistant Lightkeeper (on probation) when a vacancy arises. If no vacancy has arisen after 12 months from the date of first appointment, a Supernumerary Lightkeeper, subject to satisfactory service, will be granted the pay and increments of an Assistant Lightkeeper. If a Supernumerary Lightkeeper fails to give satisfaction during the training period his appointment may be terminated.

```
A Supernumerary Lightkeeper is not ordinarily permitted to
have his wife residing with him at a Lighthouse.
When reporting on the training of a Supernumerary
Lightkeeper, the Principal Lightkeeper must take care to
give particulars of any failings which in his opinion
render, or are likely to render, the Supernumerary unfit
in any respect for service as a Lightkeeper. It is of the
utmost importance that only suitably qualified persons
should be admitted to permanent employment
```

... It was important then that I create a good impression in order to progress in what I now deemed my chosen career. There was one piece of good advice my father gave me before I joined the Army; "Son", he said, "If you create a good impression in the first few months it will last throughout". It proved to be good advice then, so I hoped to follow it now.

Although not said in exact terms the suitability for someone becoming a Lightkeeper lies not just in their abilities so much as their character and I believe the incident reported to have taken place in the 1950's at Little-Ross Lighthouse near Kirkcudbright, where one Lightkeeper had used a firearm against another had a great bearing on the inclusion of this section.

Chapter II
Killantringan Lighthouse

So I was to be a spare man until such time as a position of Assistant Lightkeeper became available and I was deemed suitable. It was with these thoughts in mind that I now became aware of the train slowing down on its approach to Stranraer Station. The day had been long and already there were signs of approaching dusk; cars and lorries still without the need for headlights were nevertheless guided by street lights which came on either by sensor or timer, a street here a street there but never all at once together.

I got into the first of the four Taxis ranked outside the station, ordering the driver to take me the seven or so miles to the lighthouse.

A signpost with Killantringan Lighthouse showed that I was not far from my journeys end, passing by a farm yard and cresting a hill I could see the sweeping beam of the light cutting swathe across the sea, the beam suddenly disappearing on touching land, and then reappearing as land met sea; continuing the rotation

Passing through a set of gates the taxi pulled to a stop in a courtyard fenced in on two sides by buildings and shoulder height walls on the other two sides. The most prominent building was the one housing the Light tower, the other I took to be residential because a figure casually dressed was coming down the steps to greet me. It was the Principal Lightkeeper, who after shaking my hand in welcome and introducing himself as Bill Rosie; the name coincided as that of my contact. He paid off the driver, (something I had not expected), and addressing me, said, "I'll put it in along with other Station expenses at the end of the month." He guided me with my bag to a bothy at the end of the engine and tower block.

"Put your bag down here and come back to the house, we've made you some supper."

This was going to be an evening full of surprises, another being; I did not expect to see a man in his early forties, as Principal Lightkeeper. I rather expected to see a character somewhat akin to Para Handy; the skipper of the Vital Spark, weather beaten and aged with time spent close to the sea.

As I thought it was late, I did not want to fall into my usual habit of rabbiting on, so I kept conversation down to just answering any questions asked of me. It could have given the impression that I preferred my own company or was just plain shy, however it would go a long way to answer why, for the rest of my free time, I seemed to be left very much to my own devices.

The meal was pleasant and both it and the company were fortifying. It was good to see Bill in his family environment, welcoming this tired and now amply fed stranger into their home, something I'm sure they had done many times before. On this occasion it was I being truly thankful.

I was not expected to start duty straight away and being truly A SPARE MAN at the Station, I was to be introduced to all the aspects of Lightkeeping at Killantringan over the next few days under the guidance and tutorship of either Bill or the Keeper on watch. I would not be required for duty till the morning and it was with some relief that this tired Supernumerary found his way back to the bothy.

The bothy had the two rooms, one a toilet with wash hand basin, the other the living and sleeping quarters. On one wall a unit housing a kitchen sink, cupboards and work top, on top of which stood a microwave and next to the unit a fridge. Underneath the sole window of the room was a table just big enough to sit two people. The other pieces of furniture in the room included a wardrobe a comfy chair and last but most importantly, bunk beds. The last time I'd slept in a bunk bed was when I was a child, however the bottom bunk looked inviting and after some ablutions, I made my way to it for some, peaceful and rejuvenating sleep.

So it was, until the early hours of the following morning. My reaction on hearing a foghorn for the first time was thankfully just one of thought and not action.

I'm sure I would have died with embarrassment having to change my newly laundered bed linen and to explain the cut to my head that would have been a justifiable reaction to such an awakening from a deep slumber. Military training and practice saved me however, teaching me to evaluate before acting to sudden changes.

The haunting sound was deep and melancholy. Three blasts in quick succession then quiet, then repetition. It was 4.30 am and I was now very much awake, the prospect of returning to sleep seemed remote, so I decided to wash get dressed and have a look outside.

On venturing outside, the fog was thick enough not to be able to see much beyond the courtyard, I was not familiar with my surroundings as yet and I felt it would be foolhardy to explore any further than I could see. Apart from the bellowing of the foghorn, there was also the thrum of heavy engines, surprising that it was only now that I was aware them, even more so considering the bothy was attached to the same building.

There seemed to be no one else about, although the Keeper on watch was bound to be somewhere, probably in the Light tower. Even in such thick fog the Light's beam could still be seen on its endless patrol, only it looked more diffused now.

I didn't want my presence to be known; for fear of showing my newness to the service that an event such as the foghorn sounding should wake me from my slumbers. I went back to the bothy to read one of the books a previous occupant had kindly left.

I must have dozed off reading the book because it was not the fog horn that woke me but my travelling alarm clock that I had set for half seven. There must have come a point when the sound of the horn had become merely background noise. I was aware now that the foghorn had ceased, so the fog must have lifted.

There was a supply of tea, coffee and sugar, with some still fresh milk in the fridge.
A note pinned above the fridge showing that all guests to the bothy were welcome to share whatever was there so long as they were replenished for the next guest. As I had neither the time nor the foresight to buy provisions, the packet of cereal found in one of the cupboards was welcome indeed. As soon as I could, I would need to go to the shops.

I was to meet Bill in the engine room at nine o'clock, so I still had 45 minutes in which to take in the surroundings. With the fog now gone, I should be able to get my relative bearings. I could now in daylight appreciate the care Bill and the other Keepers took in maintaining the Station. The buildings and the boundary walls were painted a snowy white; the corner stone's of the buildings painted sandstone to add contrast and intensify the whiteness. Standing sentinel over all was the Light tower. The white of the column was broken only by staggered windows around its girth, and it lead up to the balcony with its handrail. The white of the column gave way to sandstone as stonework gave way to metal of the Lightroom and upwards again to the triangular shaped glass panels of the Lanternroom and finally the black of the dome topped by a weather vane.

There were other outbuildings dotted about and all were painted the same snow white, every door painted a dark green. Behind the engine room I could see a large square tank with three upright cylindrical tanks and some pipe work all painted in a shade of red I had not seen before ... I later discovered paint cans with a picture of the Forth Rail Bridge and the words Red Oxide Primer on them; indeed I was also informed that at the time the Lighthouse service were the only other customers who bought it...

Looking seaward over the boundary wall I could see that the Station was only some 60 to 70 feet above the sea, to my left there appeared to be a cove or inlet because the coastline disappeared then re-appearing a half mile or so further down, on the point where it disappeared there was a short square tower with a horn on top with two of the same red tanks but this time they lay horizontal. The tower was painted the same snowy white and the horn painted red. There would be no prizes for guessing that it was the foghorn.

To my right the coastline appeared to be relatively straight, and without visible landmarks it was difficult to know how far I was actually seeing.

On the horizon I could just make out the shapes of two ships, when I say horizon I should say the point at which sea and sky converge, because the fog or mist of earlier had not entirely lifted. I was sure that given clearer conditions the landmass of Ireland would be visible, as it was only 18 miles from this point.

Footsteps behind me announced the approach of someone, the questioning voice saying; "Sleep Well?" assured me that it was Bill. It may well have been Bill on watch earlier, so I didn't want to get caught in a lie, I replied;" Yes up until the time the fog signal started." With a smile and what I could faintly detect as a chuckle, he said. "You'll get used to it, in time." Whether or not he believed me when I told him that after a while I had fallen back to sleep he kept to himself and he never mentioned or had the report of anyone seeing me earlier.

He guided me through the double doors of the engine room. Two other keepers were waiting, both dressed in overalls and looking very much as though they expected work of some kind was on the agenda. I was introduced to Duncan and then to David. Duncan was the 1st Assistant and he would be going on holiday in three days time. I would be taking over his watches from then on until his return in a fortnight's time. Bill would give me training till he was satisfied I was capable of standing watch alone. Duncan's comment on that was, "You had better be able to stand watch or my leave will be cancelled."

There didn't appear to be threat or consternation in his tone, his comment merely to emphasise the possibility. Bill said, "Come on Duncan, It's not Rocket Science," hoping to settle the discussion and to give me a little reassurance.

David was the Local Assistant and he lived in Portpatrick. He had been on watch since six o'clock so it was probably him who turned the fog signal off. He would be on watch till midday, Duncan would then take over and his watch would be from 12 till 6. Bill and I would then relieve Duncan.

In the meantime until midday, Duncan and David would clean the Lanternroom, Lightroom and stairwell. If everything in those rooms was as immaculate as this engine room, I could not imagine there would be much for them to do.

The tiled floor was spotless with three large engines taking up the space of the bulk. Each standing almost as tall as me but longer in length were mounted on a base and painted the same green as the doors. Gleaming pipes and brass work emanating from each carcass. At the front of the engine in the middle was a brass tray with copper jugs of different sizes, each as shiny as the rest of the brass work. There seemed to be brass everywhere; encasing gauges, valves and dials. Not one piece bore the traces of neglect.

There was still heat coming from both outside engines, so I deduced that they were the ones powering the foghorn.

"I'm going to train you how to start and maintain the fog signal," Bill said.

"But first you must be able to recognise when we need to sound the horn," he continued.

Taking me outside and to the section of the boundary wall where I was standing earlier, he pointed to a headland on the horizon and beyond the arm of cove and said, "That headland there is just under 13 miles away, if that disappears, you should be thinking about sounding the horn", adding; "and you see the outlines of that clump of trees, that's just before Port Patrick, if that disappears you should have sounded the horn ten minutes earlier." Pointing in the other direction, "The furthest point up the coast you can see is about four miles, if that starts to disappear, sound the horn," and finishing with; "If you are in any doubt, sound the horn."

Confident that I need not doubt my judgment, we went back to the engine room.

On our return Bill said, "I'm glad we had that spot of fog this morning, it gives me chance to show you how to warm start the engines." It turned out that if there was sufficient heat in the engines from a previous running, all you need do is press the starter motor switch and throw a lever by the cylinders, forward, which Bill now duly did, on the engine to the right of the door which he named No 1. I hope you were not expecting something more personal like Daisy or Penelope, if you were, I'm sorry to disappoint you. For the record of our monthly correspondence all the engines at every station were from right to left of the engine room door, numbered; 1, 2, 3...

The engine responded by coming to life, the heavy thrum increasing as adjustment to the throttle lever allowed more diesel fuel into the injectors of the three large cylinder heads.

With the engine now running smoothly, he turned his attention to the middle engine of the three. "This engine," he said; "Is too cold to start in this way." He proceeded to go to the back of the engine where there was a tray similar to the one at the front. On this one though there was a brass cylindrical container with a screw cap and spout and next to it, was what could only be described as a brass insecticide sprayer. Both had the distinctive smell of petrol about them. As we were in an engine room they obviously having nothing to do with horticulture. Passing the sprayer to me to hold onto, he took the cylinder to a curved cast upturned cup at the top and front end of the engine. Tipping the cylinder into the cup a measured amount of petrol flowed into the cup. Exchanging the cylinder for the sprayer, he opened three taps at the side of the top of the cylinder heads and injected two doses of petrol in each. When he'd finished he passed it back for me to return to the tray along with the cylinder.

The lever which he had thrown forward on engine number one he ensured stayed in the back position before pressing the starter motor, as with engine number one, engine number two burst into life, after a few seconds he threw the lever forward. Initially the engine slowed but then began to pick up as the flow of diesel now combusted in the cylinder head.

This was the first time I had ever encountered an engine that started on petrol but ran on diesel.

With both engines running and their throttles adjusted, Bill drew my attention to the double cylinder shaped casing at the rear of the engines. With the engines running it was difficult to hear what Bill was saying but I was; by his gestures and the occasional word or two, able to pick up the instructions and message he was trying to convey.

The engines powered compressors supplying air to the three large red receivers outside ... the cylindrical shaped tanks I'd noted earlier ... they in turn fed air down to the fog horn through a large bore pipe. The control of the air being governed by a series of valves, the first of the valves was on top of the compressor; two brass horizontal wheels which when turned anti-clockwise released compressed air into the system, travelling up a three inch pipe to the large nine inch pipe above

The next horizontal valve was a larger wheel at the base of the large pipe, which again when turned anti-clockwise allowed air to the receivers outside. Finally a vertical valve of the same size allowed air from the receivers down to the foghorn.

Bill stressed the importance that when closing down the foghorn, the valve wheels must be closed tight in order to retain the pressure in the receivers. Otherwise it would take quite a time for the receivers to get up to operational pressure.

When the vertical valve was opened and the air travelled the hundred or so yards down to the fog horn a few seconds later, we were assured that our efforts were rewarded and all was well as the sound of the horn came back to us. After two sequences of three blasts, Bill closed the foghorn down, in the reverse order of valves. Closing down both engines by pulling the throttle lever up and setting the lever by the cylinder to a vertical position, which I assumed must prevent compression causing the engine to stop.

After he was happy that I was able to start the engines and could operate the fog signal he said; "Anything you are not certain of"?

There was nothing in what I had been shown that was difficult but there was something that did intrigue me and so demanded answering. "Yes there is", "Is that a starting handle at the front of the engines?" Bill contemplated for a second or two before answering; "Yes it is," "but you'll regret the day you ever have to use it," then saying "Go on try to turn it."

Feeling brave and knowing that Bill would not be satisfied with me just taking his word for it, I rolled up my sleeves and took hold of the handle. I had to adjust my stance a few times before I could get enough leverage on the handle to move it at all. Not having any success with a bellowing laugh Bill said, "It might help if you remember to make sure the cylinder lever is vertical, you'll never turn the engine over with the compression on." "Oh and by the way, this is a task that can only be done with two people." "You turn the handle and I'll throw the lever forward when there's enough impetus."

Even with my mistake now corrected it took all my strength to turn the handle. At the end of my strength now, I was overwhelmingly relieved to hear the engine kick into life. "Well done" Bill exclaimed, "But imagine having to do that to an engine that is cold." The thought of it would dwell long after I recovered from the effort.

I was now recovered enough for Bill to take me down to the horn itself, but not before he let Duncan know where we were going. It was much wiser to be instructed on the mechanism of the foghorn without the horn being on. At close range even earphones would struggle to protect you from the sound and the vibrations could also damage a person's internal workings if exposed for too long a period.

We followed a well-worn path down to the foghorn. We could track the large bore pipe going into the side of the receivers and from the second one into the side of the white painted building with the horn protruding from the top.

The horn was not quite big enough for a man to crawl into but certainly big enough to hide a child. Again it was painted the same colour of red. The small room inside was barely big enough to take the two of us and had just enough headroom for us to stand. Bill was a few inches shorter than I so was comfortable, I was in fear of banging my head on the enclosed light barely an inch above, which good sense and the dimness of the interior had caused to Bill to switch on.

I could see now where the end of the horn came through the ceiling, and entered a large drum shaped housing, below this was a bronze conical frame that given its dimensions and the fact that there was a toothed hoist which could raise and lower it into position, must have had some considerable weight.

This I was told was the reed, raising the hoist to support the reed and releasing the clamps that held the reed tight into the housing, he gently lowered the reed for me to see. It was made of brass with a horizontal rotating disc with angled vanes, mounted on top of a supporting frame. A fine film of grease or other lubricant ensuring perfect freedom of movement.... This was the only piece of brass never polished at the station ... It was cleaned with great care to remove any dust or debris and then re-oiled lightly with just enough to ensure movement not enough to endanger the function by attracting dust. He inspected the reed and was satisfied that it was not in need of cleaning or lubrication he replaced the reed.

Coming from the side of the housing was a pipe leading back to what I was informed was the timing mechanism that gave the horn its characteristic three blasts every minute. The clockwork like mechanism was started by air fed into the side from a pipe that led back to a valve and then to the main large bore pipe

The narrowing from large to small pipe increased the air pressure. When air entered the mechanism, it started a pall on a cog which had cams attached at intervals, when the cog rotated a cam would depress a valve button, thus releasing high pressure air to the reed starting it to rotate at great speed, the next cam would allow greater air pressure into the reed causing it to vibrate. The horn amplifies the sound in the same way the horn of the old gramophone does.

I had been tutored into the workings of the fog horn and we would now be returning to the engine room for some instruction on maintenance and cleaning of the engines but not before we had a tea break.

Any repairs that required attention were the province and the office of artificers who came to the station twice annually or when a malfunction occurred. This applied to all machinery electrical or mechanical at the station.

Our duties were to make sure that the machinery had enough oil and lubrication, this meant checking the dipstick and filling the oil reservoirs at the top of each cylinder. Water was fed to the engines via pipes coming from the large square tank outside, opening a tap on the engine would ensure that this flowed freely. With the engines now checked, it was just a matter of cleaning the brass work that had acquired our finger prints and give the engine room a general clean, the workshop and the oil store off the engine room were also given a clean and tidy up.

With the appearance of Duncan and David and a glance at his watch, Bill dismissed them and told me to meet him here at six o'clock. Duncan was now on watch so he was not going to leave the station, he had only a short walk to his house below Bill's in the accommodation block and it was there that he and Bill were going. David was heading towards his car parked just outside the gate.

If I had the presence of mind, I would have asked him for a lift into Portpatrick. As it was I gave acknowledgement to my instructions and bade them goodbye, making my way back to the bothy.

This was just the chance I needed to get some provisions and after a cup of tea, I duly set off for Portpatrick. I would be retracing the route backwards of the evening before except this time I was on foot. I did note that I had to turn right at the road junction. There a mile stone confirmed I was heading in the right direction and at a distance of three and a half miles lay Portpatrick, not far for an ex-soldier like me...

It was during my leave when Margaret and I had met for the first time.

One sunny Sunday morning she said, “Let's go on a picnic, it's such a lovely day, we could walk to the Castle at Loch Doon.” Not knowing Margaret well enough to know whether this was an order or request and thinking it would be a great opportunity for us to be alone, I replied, “Great Idea” and said it with enough enthusiasm to sound convincing.

What did give me some concern though was I had not bought with me shoes suitable for walking too great a distance, platforms were the height of fashion at the time, so it was with this in mind that I questioned, “Is it far?”, I said inquisitively, trying not to sound as if I wished to put a dampener on the venture. “No, just a couple of miles” was the reply.

Well I defined a couple as being two and thinking, doubling the two to four for the return journey didn't sound too bad and I thought I should manage it even in the platforms. It turned out to be five miles just to the Loch and the wear on my heels was already starting to draw blood, the soles seemed to be generating enough heat to boil a kettle.

The Castle was as yet nowhere to be seen so I suggested we stop here for lunch. The scenery was beautiful, but at this moment my thoughts were to finding some relief for my feet. While Margaret prepared the picnic, I went down to the water's edge, taking off my shoes and gingerly removing my socks that I could see were worn to holes and bloody at the heels. The initial shock of the ice-cold water and the sting as it hit the wounds was soon followed by euphoria as the pain subsided.

I was beginning to hope that if we stayed here long enough, Margaret would forget about the castle and we would make our way back home.

I tried without making it too obvious to delay lunch as much as possible and just when I thought I might succeed; she asked "Are you ready? Then let's go to the castle". Looking all around and hoping beyond all hope that by some miracle the castle had either been destroyed without trace or just up and vanished, I asked, "Where is it?" Pointing to a clump of trees that I estimated to be about two miles away, she replied "Just around the bend beyond those trees".

Once more braving the trek and thinking it might be less painful if I walked in stocking feet, and then having that thought evaporate when I remembered about the holes, we marched on.

When we got to those trees and there was no sign of the castle I thought God had answered my prayer somewhat belatedly, however, before I could raise the question again, Margaret in an apologetic tone, said "Sorry it's not this clump of trees but the next one". That clump of trees appearing to be even further than the first. If Margaret could have read my mind at that exact moment I have the sneaky suspicion that our futures would have taken an altogether different path.

The thought passed and I resigned myself now to getting to this castle no matter what, disappointment turning to determination. The disappointment returned with a vengeance when we eventually got to the castle. I expected to see a sight befitting our endeavours and the equal of Stirling or Edinburgh. What lay before my eyes was an incomplete structure; barely four walls, no roof. A ruin!

It's funny what we do and suffer in the name of love, but I believe it is things like that make the bonds between us stronger. To be sure there were good lessons learned. As soon as I had bought replacements, the fashionable platforms went in the bin and never again would I take Margaret's word when she began a sentence. "It's only".

It was with shoes better suited for walking and legs made accustomed to middle distance hiking that I made this quite pleasant amble into town. Nothing was outstanding or remarkable about the trip there or the return journey except to say that the way back was made arduous by the shopping I carried.

Portpatrick was typical of a small harbour town, the crescent shaped sweep of the cove with terracing on the hillside and houses dotted here and there. A large Hotel atop the hillside on the right overlooked the harbour below that sheltered a scattering of craft within. An avenue of shops and businesses came down the hill to the promenade, finishing with a row of cottages each painted in contrast to its neighbour reminiscent of the picture postcard frontage of Tobermoray. I would make a few more trips into Portpatrick and on each trip find something new to see before my time at Killantringan came an end. My mission today was to buy food, so I would discover more of Portpatrick on another occasion and think about taking an alternative route back.

I was still a little weary when I returned and welcomed the prospect of making myself something to eat and settling down to relax before meeting Bill for the watch at six. As there was no TV, I would just have to continue reading the book I had started. This was a pastime I grew very fond of during my time as a Lightkeeper, even when the opportunity to watch television presented itself.

With the evenings of early April lengthening each night, it would be past eight before sunset slipped into dusk and then into the darkness of night, Bill however had the one or two things to show me before we needed to light up.

With no indication that the prospect of fog was imminent, the sky was almost clear with only a few fluffy clouds here and there and visibility very good. I could now see the coastline of Northern Ireland on the horizon. Bill and I climbed the staircase up to the Lightroom.

The efforts of Duncan and David were evident as not one speck of dust could be seen and the paintwork was as fresh as if it were newly painted. The treads and risers of the stairs painted a vivid Post Office red as opposed to the red of the receivers outside. The walls painted buttermilk and the window insets a pale blue, the windows white. The insets must have been nearly two feet in depth, indicating the thickness of the walls of the tower. Here is a structure that is built to last, I thought.

The open stairwell with its varnished banister; spiralled some 40 feet or so to a landing, then a steeper cast metal spiral ladder with polished brass hand rail led to the Lightroom above. A weight painted the same blue as the insets hung suspended by a cable coming through the floor of the Lightroom.

The Lightroom was painted a biscuit colour and the walls were windowless, the apertures I had seen from below the night before were for ventilation. On the opposite side of this rotunda was a door that led out onto the balcony and just before it was a metal ladder leading up to the Lanternroom. Next to the door was a gray electrical cabinet, fed by a cable entering the bottom; leading out from the top was another cable that travelled up the wall to the platform grating of the Lanternroom.

The most dominant feature of the Lightroom was what looked like an enormous carriage clock; standing almost my height and nearly four feet square. The frame was of cast metal of some kind, painted in the same blue. Glass panels in the sides allowed the clockwork mechanism to be visible. One side of the casing had glass doors for access. A handle finished in brass, protruded from the side to the left of the doors. At the top of the casing was a column supporting an equally enormous plate on which roller bearings ... like the ones that support the platform for the cars of a waltzer in a fairground ... in their turn they supported the huge weight of the lens carriage.

On a desk next to the gray cabinet was a book opened to reveal the names addresses and comments of visitors to the lighthouse, the desk certainly had some age to it, given the patina it had acquired and the inkwell set in the top long since passed usage. The top was also hinged and the contents within its compartment as yet unknown. Next to the desk sat a chair. Above the desk was a timetable indicating the times of sunset throughout the year ... This chart held the most critical piece of information needed by the keeper on watch because sunset was the official time of lighting up. Lighting up any later would be deemed a serious breach of discipline and could result in dismissal...

With sunset not too far away we made our way up the ladder to the Lanternroom. The impressive sight of the lenses overwhelmed everything that I had seen so far. The platform in which we stood was about 30 inches wide and went the circumference of the Lanternroom; the lens carriage dominated the rest of the breadth. The lenses, for there were two of them, were sets of angled glass prisms set into a brass framework tapering outwards from the top down to a central lens, then tapering inwards again to the base. The lens was like the lens of a seriously myopic set of glasses but on a much larger scale. The two lenses did not occupy the full circumference of the carriage; the remaining section was a wall of prisms from top to base; hinged at the middle so that it could be opened up allowing access to the centre. In the centre was a pedestal on to which a pear shaped light bulb was screwed.

This was a mercury vapour lamp and was the source of the light. I cannot remember the exact wattage but think only about 500 watts and it was about twice the size of a domestic light bulb. It was the lenses that intensified the light to around 1,000,000 candelas ... one candela being the approximate power of the light emitted by one candle... It was the tapering of the prisms that focused and concentrated the light towards the bull's eye of the lens. When seen from the sea and at a distance it would appear the light was flashing twice in succession every minute.

The framework or astragals of the lantern were made of a series of triangular shaped panels three rows high and plated with glass; this was topped by the dome. Both astragals and dome painted black.

Outside you could just make out the outlines of a ledge. The landward side of the lanternroom had been painted out ... this was the reason of the sudden disappearance of the beam I had noticed last night ... Just so it didn't blind or cause discomfort to local residents or passing road traffic. After all they did not have the need of a navigational light or pay for the services of one.

Around the inside of the lanternroom hung net curtains supported by a rail. Bill took one side of the partition and gestured me to take the other. We drew each side back to the blanked out section. "It's important that you remember to draw these curtains in the mornings when you have put the light out" Bill said, then went on to stress, "If you don't, the lens is powerful enough to act as a magnifying glass, scorching anything combustible that it focuses on," pointing to a set of wooden steps that bore the tell tale marks of such an event.

We returned to the Lightroom and Bill opened the front of the casing. Inside was the mechanism whose function was identical to that of a clock, in this instance not providing power to drive hands but to drive the lens carriage. Because most of the parts were of brass it received the same care as the rest. There was a small lever Bill moved which disengaged a pall from the cog wheel at the base of the lens carriage. Moving to the lens carriage he pushed it lightly just to start the momentum, the machine taking up the pace.

There came the continuous ... Ding, ding, ding sound of a bell; letting the keeper on watch know that the machine was in motion. There was a brass circular dial with the words slow on the left and fast on the right and graduations marked around it. A brass indicator was set just past the half way point, airing towards fast ... By adjusting this needle you could make the carriage go faster or slower, you could also adjust the speed of the carriage by adjusting vanes on the flywheel. Adjustments to either would have a serious affect on the speed of the carriage and therefore the characteristic of the light. Because this was a difficult and delicate operation it was only to be carried out by the Principal Lightkeeper. I was to be very careful that I did not disturb either when operating the machine...

With the lens carriage now rotating, he closed the front of the case. He then moved towards the gray electric cabinet on the wall, turned the largest switch to on, and looked up through the lens to see that the light was indeed on. "Well" he said, "It's as simple as that, any questions?" I wasn't able to think of any at that moment but I was sure that there might be later on.

Bill thought that this would be a good time for me to fill in my details in the visitor's book, before doing so he lifted the lid of the desk and produced a stopwatch. "You might need this to time the light." Handing it to me adding, "all good keepers should time the light at least once during their watch," he then pointed to a mark on the carriage as it came round; indicating where I should start. "But don't time the light till after the first wind" he said; pointing to the handle on the side of the casing, he continued with, "It takes 40 minutes for the weight to drop to the bottom of the tower, so don't go wandering off" then,

"I don't ever want to hear the alarm bell ring because you've forgotten to wind up." After he'd satisfied himself that all was well in the Lightroom we made our way back down the stairs. At the base of the tower on an arm in direct line with the hanging weight was a switch lever that would set off an alarm should the weight touch it.

The weight had already dropped a foot or so and I could now see that on top of the weight was a stack of old pennies none of them were of recent mintage and all bore the signs of age. Before I could raise any questions, Bill ventured, "So delicate is the timing of the mechanism, that even pennies can alter it" and; "in winter when the temperature drops it can slow the light, adding a few coins will bring it back to time".
Bill asked me if I was happy enough to stand the watch and if so he would be back at half nine, with no objections from me Bill left.

I did remember to bring along the book I was reading but before returning to the Lightroom and settling myself down in the chair. There would only be a couple of winds till the end of the watch so I made the mental note when the first wind would be. A brass wall clock in the watch room and an occasional glance would warn me of the impending wind up time.

Grasping the handle I began to turn relieved to find the resistance less than that of the engines, it was still an effort though, winding the weight up to the top, satisfied that I had mastered this task I went down the stairs to see if the weather was still clear.

Now that it was almost fully dark, I could see the characteristic flash of other lighthouses, but as yet, not knowing their identity. Seeing them and the light of what I assumed were vessels of some kind, I was assured that visibility was good and so returned to the Lightroom. This process was repeated till Bill appeared right on the dot at half nine. "How did you find your first watch?" Bill asked. I was not sure how to answer this; as Bill so kindly put it this morning ... It's not rocket science ... coupled with the fact that I had stood watch or guard duty plenty of times before, I settled with the reply of "uneventful, thank god." Bill and I exchanged a few pleasantries before going downstairs.

The change of watch took place in the engine room, the dinging of the machinery could be heard even from down there and a glance up the stairwell was indicative that the machinery had been wound; the last act of the Lightkeeper going off watch.

Duncan duly appeared at a couple of minutes before ten, yet a few more pleasantries passed between us. My next watch would be with Bill at six in the morning and without further ceremony I made my way back to the bothy.

This then was the end of my first day as a Lightkeeper; nothing I had done was taxing either mentally or physically; nevertheless, I felt the tiredness that novelty can bring on. After completing my ablutions a light supper and a cup of tea, I set my alarm for 5.30 and settled into bed. My last thoughts before falling into a sound sleep were ... I wonder what tomorrow will bring...

Walter, the 2nd Assistant lived with his family the house next to Duncan. We met in the engine room. He and Duncan had exchanged watches to allow Walter to attend an appointment; this suited Duncan just fine, because it would give him the equivalent of an extra day's holiday. It did strike me as odd that Duncan was on watch yesterday afternoon then again at ten, making it difficult for me to work out a roster. Bill would be giving me a roster later on this morning, so all would become clear.

Bill entered the engine room just before six o'clock, like the previous evening a few pleasantries were exchanged and then Walter went home...

This was to be the cycle of events at the change of watch, every day till my departure; everyone had their own lives to live when they were off watch. When you think about it, it was not really much different from the normal 9 to 5, the main difference being when we were at stations like Killantringan; our homes and place of work were the same place. It's not that the Keepers didn't socialise together; I had seen the keepers wives talking as neighbours do and I had also seen them dressed up to go to what I assumed was the same venue. I did not feel in any way excluded from their circle and I accepted and respected their wish for privacy as they did mine. It was a struggle though not having someone to talk to and this is where I missed having Margaret's companionship. I have never found it easy to make lasting friendships and the number of friends can be counted on one hand. Of course there have been acquaintances but seldom have they lasted beyond the moment. Margaret then was not only a wife but also my confidant and friend. Our conversations on the telephone while I was away had that muted quality about them as though someone else might be listening, so we wrote to each other, just as in the beginning although not quite so profusely; at least one letter a week till my last week ...

Sunrise was fast approaching; heralding the prospect of a fine day so Bill and I made our way up to the Lightroom. "Can you think what we need to do to put the light off?" he asked wandering perhaps if I had forgotten the sequence at lighting up time. "Yes I think so," I replied. He then went on to say, "Have a go, and don't worry if you do anything wrong I'll soon put you right."

Hoping that the sequence would be in the reverse order, I started by turning the light's switch to off," Good, well remembered", he said. I then applied the brake to the lens carriage, and once the carriage had stopped rotating I went up to the Lanternroom and drew the curtains. My final task was to wind up the machine. "Nothing to it, is there?" He said approvingly, not feeling that the remark required an answer, I merely nodded in acknowledgement. As today was Saturday there would be no turn out for work, there would just be Bill and I on fog watch till noon.

He handed me a sheet of paper with my watches for the next two weeks till Duncan returned; when once again I would be The Spare Man at the station. My first solo watch would be at 6 am till 12 pm on Monday morning, the roster then following the pattern; 6 am - 12 pm, 10 pm - 2 am, 12 pm - 6 pm, 2 am - 6 am, 6 pm -10 pm then a day off, then repeating itself.

Bill was happy that so far I had acquitted myself well and he had no reservations when it came to watch keeping, there was still quite a lot to learn, most of the things would only be relevant if certain situations arose and he thought that I possessed enough common sense to be able to deal with them when the time came.

He asked me if I minded taking the watch till ten o'clock so that he could attend to something, as I had assumed that I would be on watch till noon anyway, the thought of manning the watch solo a reassurance in his confidence of my abilities.

I now had the task of finding things to do to occupy me for the rest of the day, the whole of tomorrow and for the rest of my time at Killantringan.

... Extract from the Regulations...

Section B Routine Duties

B.1. CHANGING WATCH.- A Lightkeeper on duty must remain on Watch until relieved. The relieving Lightkeeper, before taking charge, is to satisfy himself that the light is efficient, that the machine is wound up, and that everything is in order. A Lightkeeper going off Lightroom or Engine room watch must advise the relieving Lightkeeper of anything unusual that may have occurred during his Watch.

B.2. CLEANING.- The optical apparatus, Lantern panes and Lightroom machinery are to be thoroughly cleaned every day, particular attention being paid to polishing the glass and bright work. If the glass is greasy it is to be washed with methylated spirits or approved detergent before polishing. When cleaning in the Lightroom has been done everything should be left in readiness for lighting.

Fog signal, Radio, Radio Beacon and Radar Beacon machinery and equipment are to be cleaned and polished frequently so as to maintain a standard of cleanliness equal to that of the Lighting apparatus. Machinery should never be cleaned while it is in motion; if it is operated electrically it must be isolated before cleaning is commenced.
The Principal Lightkeeper will be responsible for the rotation and the apportioning of the day's work among the Lightkeepers and for its efficient execution; he should assist in the work himself, so far as his other duties permit.
The tower stairs passages Engine and Radio rooms, stores, doors, windows, paths, lawns, grounds, etc., are to be cleaned and tidied whenever considered necessary by the Principal Lightkeeper. Cleaning should normally begin no later than 9a.m., during the months March to September inclusive, and not later than 10 a.m., during the other months of the year.
The daily routine cleaning as described should as a general rule be confined to an average of two hours. At Three-man Stations the Lightkeeper coming off watch at 6 a.m., need not normally participate in the above cleaning work during that morning throughout the year. At Two-man Stations, the Lightkeeper coming off watch at 6 a.m. may be excused cleaning work during the winter period only.

For certain seasonal work such as lime-washing and painting which needs to be done while fine weather lasts or when there may be a risk to personnel when working short-handed, the Principal Lightkeeper has full discretion to call upon the morning watchman to assist if required.

B.3. EFFICIENCY OF THE LIGHT.- Constant vigilance and strict attention to all the details of Lightroom duty are required of Lightkeepers in order that the light may be kept at its maximum efficiency. The ventilation should be carefully regulated to prevent condensation on lantern panes. If the lantern panes become obscured by spray, snow or any other cause they should be cleaned as soon as possible.
A spare lamp, vaporiser and mantle complete are to be kept ready for immediate use when necessary. Where the navigation light is electric, defective electric lamps are to be replaced immediately, unless otherwise instructed.
The emergency lamp is always to be kept ready for use, it should be tested once a month, or once a week if electric and a note to that effect entered in the Monthly Return.
If the emergency lamp has had to be used the Principal Lightkeeper is to report the fact to Headquarters by the quickest means and state the period of time the Light was defective and for how long the emergency lamp was in use.

The emergency light at Killantringan at that time was what is known as a Liverpool Lamp very similar on operation to a hurricane lamp. If the electric power had failed we would have had to light this lamp and then remove the mercury light bulb and then place the Liverpool Lamp in its stead, if however it was just the mercury vapour light bulb that had blown, we would just change the bulb so therefore returning the light to its maximum efficiency sooner. The light offered by the Liverpool Lamp was far less than the mercury vapour. A standby generator was due to be installed later on that year so negating the need for the Liverpool Lamp altogether. The lamp was stored ready for use in a cabinet by the stairs to Lanternroom.

©Scottish Lighthouse Museum

Chapter III
At a loss

Painting at Killantringan was due to start once Duncan returned from his leave, I would not be asked to help because all the painting was done and paid for on a contract basis, keepers were paid so much for a window, a door, etc so it would be difficult if not impossible to allocate work fairly or ensure that keepers' had the right payment for work they had done without the added burden of having additional staff on temporary assignment. At Stations like Killantringan the general consensus was; that when all the works orders for painting had been completed for the year, all monies owed would be shared equally among Keepers at that Station. So that was something I would not be able to do not do to pass my time.

I have always been keen on keeping active in some way or another, but this was the first time I had been away from my home environment without my usual and familiar pastimes to rely on. I decided that I would revisit Portpatrick tomorrow and take an alternate way back, but this afternoon I would do a bit of beach combing to see what I could find, checking out the local vicinity in the process.

There was a well-equipped workshop at the Station so there would be the opportunity to make something of whatever I might find. I had no definite plans in mind but I would allow my creative inspirations to flow from whatever fortune was found in my path.

The coastline immediately below the lighthouse looked bereft of flotsam and jetsam; even if I were to see anything worthy or notable, getting to it would be hazardous retrieving it and bringing it back suicidal. I would have to start my search to the left of the foghorn where the land had seemed to disappear. I followed the path to the fog horn then veered left, as the land started to fall away I could see now that there was indeed a shallow cove.

The terrain sloped gently towards the foot of the cove, but where I was standing at the mouth of the cove; the slope was steeper, ending abruptly at the cliff edge. The grass was knee deep, and I hoped that I would not disturb an adder on my expedition. The other arm of the cove had trees sweeping back into more general woodland inland. The trees were tall and bereft of branches at low level, the canopy a lush dark green, not being educated in arboriculture; I could only assume that they were firs of some kind, which had been planted by man. Their regimented outlines broken by the light that passed between them, not enough gaps to indicate what lie beyond but enough to show that wood had little depth to it.

The cove was not more than a couple of hundred yards from arm to arm, so I would be fortunate indeed to find anything here. There was no beach as such just layers of tennis ball sized stones all made of much the same material, none that I could find had a highly polished surface or distinctive coloration that I sought. I was not desperate enough nor indeed would I ever be, to want to return to this spot with a hammer and start cracking open the rocks in the hope of finding a fossil of some kind.

I was about to give up and move on, when something caught my eye just below the surface at the waterline. It was about six inches in length and about three inches in diameter. Lifting it out of the water it was obviously made of copper, brass or an alloy of both because of the thick layer of Verdi Gris encasing it. Now my creative mind swung into action, admittedly I could have spent more time thinking of something more innovative but perhaps less appropriate given the circumstances.

I had the makings of a lighthouse in my hands, and it would neither take the effort or the technical skills to transform it. What I was looking for now was something suitable to make a base for it to stand on. The chances of finding something made in copper or brass of the dimensions that might do it justice, seemed far too remote. It turned out that I did not find anything else of use so would continue my search at a later date.

I took the lump of metal back to the workshop. Bill had already given me permission to use the tools there; however I was banned from using the powered lathe.

Sitting in a corner of the workshop stood a treadle lathe; long since used, but free from dust. On the shelf above were a set of tools not rusty but bearing the patina of time without use, these tools designed specifically for working metal. Underneath the shelf in a rack were chisels for working wood. All could do with sharpening. The workshop was equipped with files and sharpening stones, so there would not be a problem there. Trying the treadle and noting the response, I could see that with a little cleaning, oiling and sharpening, I would be in business.

The plan was to use the lathe to clean and dress the column, then taper it slightly, I would round the top and use gouges to form the windows, astragals, stonework and doors. Having cleaned off the Verdi Gris, I could now see that the column was solid brass ... Where it originally came from and how long it had been in the cove would be for some other detective to deduce, I can only suppose and Bill concurred that given the proximity to the Lighthouse, here was where it had its last use. Bill could not recognise it as having shape or form of anything used either today or in his time, however he did suggest that Artificers' made their own parts and tools in days gone by and maybe this column was the rudiment of some part of machinery that had accidentally been discarded ... whatever, for now I would endeavour to transform it into a lighthouse paperweight.

My plan to go into Portpatrick could have come to nothing if the forecasted weather had materialised, as things were it was ten o'clock before I left. The only purchase I intended to make was; newspapers and a pint of milk, the main purpose really being to get some exercise and take in the spring sunshine which had at last a little warmth to it ... You cannot really get lost if your destination lies on the same stretch of coastline, the interesting and somewhat baffling part is that; you follow a marked footpath and then that path suddenly disappears. The exact point at which this happened on my return trip will also remain a mystery because I was not to repeat it on subsequent trips back from Portpatrick along the same path. There's an old saying in Scotland, "What's before you will not go by you" and maybe I was destined to take this route, any way there were two notable things that happened by way of it...

I climbed the hill that leads from the main street up to the hotel on the cliff top. Beyond the hotel was the golf course and it was there that I picked up the footpath marked as leading to Leswalt; which was just a bit further up the coast from Killantringan. At the bottom end of the golf course the path led over a style and into pasture, the cropped grass sloped away and obscured from view its extent or possible occupants. Wary that I might encounter something slightly larger than a cow but a little more aggressive, I was happy to see the style at the other end within easy sprint distance should the need arise ... Thinking about it now; why would a farmer field a bull on land used as a footpath? ... It must have been somewhere within the woods beyond the pasture that I wondered off the path; veering left perhaps when I should have gone right, anyway, after a little while, evidence of the path became less to the point of being indistinguishable.

Thinking positively I decided that I would and should if I kept on my current bearing, pick up the coast at some point. Just in case I had entered the twilight zone or found myself completely disorientated that I might walk round in circles through this wood, I made a mark on a tree with the penknife that all good Lightkeepers should carry. The mark was not deep enough to cause injury to the tree but just enough to be noticed.

It was not long before the wood came to an end opening into pasture once more. The rise of the land to the left and the nothingness beyond an indication that the coastline was close at hand, I chose to maintain my present heading to at worst pick up the road leading back to the lighthouse or to find the cove of yesterday, the clump of trees ahead of me having a familiar appearance to them. This stand of trees was indeed similar but on passing through them it was not the same cove that opened up before me.

This cove was slightly bigger and it was sandy, furthermore, there was a brick construction with a large plate screwed on one wall with the information that this was the property of so and so and housed the cable carrying telephony across to Ireland, to bear witness to this; from the seaward facing side of the building the cable housing emerged only to disappear beneath the sand some ten feet away.

The plate was aged and the building not seemingly maintained, its only aperture; a door, had not seen paint in many years. Protected from the weather by the trees that encompassed it; and the lichens that adorned the concrete roof would have soaked up any moisture collected.

Any recognisable path leading to this structure had long since overgrown and I had no intention of finding a path that would lead back to the road. No, by far the quickest route to the lighthouse was to keep to the same bearing which would surely lead me to cove of yesterday.

It was on this tack that I noticed the gnarled edges of a flat piece of wood lying on the ground. It was about 13 inches at its widest and about two inches thick and it was hard. I struck it off the trunk of the nearest tree and it did not break. This would make the ideal base for my paperweight. I had the three things to be thankful of that day; I had come across the cable house, I'd found the base for the paperweight and I'd found another route back to the lighthouse.

The routine of watches passed without incident and my time off was occupied with the paperweight that was now almost complete. Duncan had already returned from holiday and I found myself with even more time on my hands as the watches were altered to accommodate the extra man. A letter had arrived from the Board five days before my departure. Inside were details of my next assignment and a payslip with my projected earnings to the end of the month; our payday was the 27th of each month.

The past three weeks had flown by, I had found other things to do just as a break from paperweight making and to get a bit more exercise; the lighthouse lawns needed to be cut, It was a chore that was less of a priority to the other duties and one that Bill was only too glad for me to voluntarily fulfil. He also had a plot in the garden that required turning over with the spade, fair exchange being no robbery, I agreed to do this especially when he offered to take me into Stranraer the next time we were off together, and he also offered to take me to the station on the morning of my departure.

I was looking forward to the trip into Stranraer especially because there would now be a surplus in the bank, something of a rarity for the past few years. With the extra money I bought an eternity ring for Margaret and two silver bracelets, one for each of the girls. If nothing else it might remind them if not me of this time of plenty.

I had now saved money on the return fair to Stranraer and with the help of Bill's daughter the cost of a haircut...
In those days my hair grew quickly and in order to keep it a reasonable length it required cutting at least once a month, so you can imagine that by now I would be ready for a trim. By coincidence Myrah; Bill's eldest daughter, was studying hairdressing at college. She was in need of a guinea pig; I was in need of a haircut, problem solved you would think, so long as I was brave enough.

It was not a thing for me to undertake lightly; the last time a trainee hairdresser was set loose on my hair, I was in short trousers and had to sit on a plank across the arm of the chair. He managed to relieve me not only of my hair but a chunk of my right ear as well. The owner of the establishment sought to buy me off with the presentation of an airfix model of a Lancaster. That would have been ok if on the one hand I had been old enough to build it and on the second it could actually fly. My father had taken great care in its construction only to be as disappointed as it smashed to smithereens when I test flew it from the second floor window of my bedroom. It was a forlorn hope that I would forget the ordeal as the memory and the scar remains to this day.

With my Good Samaritan spirit overriding the memory, my only remark in reference to the event; and one that she must have thought completely abstract as I did not go into detail was, "I hope you know I have outgrown airfix models."
She must have practiced all the skills she had recently learned because it took over an hour to complete. I do hope that she was as proud of the results as I.
With my bank balance in a healthy state; my hair cut, my lighthouse paperweight now finished and my spirits high, I felt ready for the next assignment, especially as I was to have the four days at home before my departure to Sule Skerry.

I like to make my own judgment about people and places so I refrained from asking Bill ... even though he came from that neck of the woods ... if he could tell me what it was like, at that moment I was happy enough to know that it was located at the western approach to the Pentland Firth that lay between the mainland and Orkney. The type of Lighthouse I was to go to was referred to as a Rock Station and I should read the special instructions on Rock Stations in the Regulations.

I had enjoyed my stay at Killantringan and would not object to serving there or at any station where Bill or Duncan was Principal. I was even happier at going home to see Margaret and the girls, especially when I had the surprise gifts for them ... As for the paperweight. The finished article gleamed like all other lighthouse paraphernalia and looked quite the part; alas it never graced a desk, and was probably lost during the many moves we made, however it did fulfil its main purpose by occupying my time.

The train journey home seemed twice as long as the journey down even though it was only half the distance. But that is always the way of it. When you are excited at the prospect of something; time has the habit of seeming to move slowly. The single-track line and the speed restrictions made things worse.

I arrived home in the mid afternoon, both girls must have been eagerly awaiting the moment; delighting in my arrival or maybe sensing that I had more in my bag than just dirty laundry, in any event I was nearly bowled off my feet with their welcome.

Margaret said very little, it was almost as if she were reverting to the shy girl of our first meeting; not sure what to say in the presence of this stranger but eventually the ice was broken by the tender hug and long kiss in welcome. And this was how it was every time we had to part and were reunited.

... Extract from regulations...

Section D. Rock Stations

The following special Regulations apply at Rock Stations:

D.1. COMMUNICATONS- (a) Radio. - The Principal Lightkeeper and his Assistants are to make themselves thoroughly acquainted with the radio routines and with the operation of the radio set on frequency of 2182 kHz for use in distress and emergency. A Lightkeeper is to be available to man the radio or telephone as the case may be at Shore Station or Control Station at the stipulated times.

(b) Visual.- The Principal Lightkeeper and his Assistants are to be familiar with the visual signals employed at the Station and are to hoist the appropriate flag signal when reliefs or stores are being landed, notwithstanding that radio communication is in use.

D.2. KITCHENS.- Cookers are to be thoroughly cleaned at least once a week to ensure that no fat or other inflammable substance is allowed to accumulate on them. Where gas cookers are fitted the greatest care is to taken to ensure that the flame does not blow out, and when the cooker is not in use the main supply valve at the cylinder is to be turned off. The calor gas alarm, if fitted, is to be tested weekly and the fact noted on the Monthly Return.

D.3. LEAVING ROCK.-Except when relieved in the normal course or in case of sudden illness or accident no Lightkeeper is to leave the Rock without permission from Headquarters and until his relief is landed on Rock.

Lightkeepers may use their own boats for fishing in the vicinity of the rock if they have the consent of the

Principal Lightkeeper, but such consent is only to be given when the weather is settled; the Principal Lightkeeper is to be informed of the general direction and distance which the boat will be taking so that a look out can be kept

D.4. LETTERS.- All official correspondence to Rock Stations will be sent in duplicate to the Shore Station, where, unless it is addressed to the Principal Lightkeeper and marked Confidential/Personal, it is to be opened and acted upon as necessary by the Lightkeeper ashore. One copy is to be retained at the Shore Station in the Station file; the other, which will be in an envelope marked "Rock" is to be sent out to the Rock at the first opportunity.

When the Principal Lightkeeper is on the Rock, the Lightkeeper ashore is to advise him at the first opportunity if the action he proposes to take or has taken.

D.5. NON-SERVICE EMPLOYEES.-Persons not in the service of the Commissioners who may be required to remain in the Rock are to be victualled in the same manner as Lightkeepers.

D.6.PRECAUTONS AGAINST ACCIDENTS.-When landing at or leaving a Rock Lighthouse, and when attending to their duties on the Rock, Lightkeepers must not needlessly incur any risk to being injured or washed away. They must acquire, by personal observation, a thorough knowledge of the sea at the landings and around the Rock generally during all kinds of weather, and note the high and low-water marks of spring and neap tides.

Lightkeepers on the Rock are to ensure that all gratings, jetties and landings are cleaned regularly and kept ready for immediate use. Life lines or additional steadying lines should be run out where considered advisable when a relief is being made in rough weather.

D.7. PROVISIONS.- The Lightkeeper on shore is required to see that an adequate supply of provisions for the Rock is ordered in good time for the relief.

The emergency stock of provisions, sufficient for 18 days at Rocks relieved by helicopter and 14 days at other Rocks and supplied by the Commissioners, must be maintained on the Rock at all times and a certificate to this effect is to be included in the monthly the Monthly Return in accordance with the instructions given in the Station Order of 29th July 1969. The Principal Lightkeeper will ensure that items in stock are kept fresh by use and replacement.

D.8. ROTATION OF LIGHTKEEPERS .- Reliefs are made every fortnight. At Rock Stations the rotation for each Lightkeeper will be four weeks' duty in the Rock followed by four weeks ashore.

D.9. SHORE STATIONS.-When at the Shore Station, The Lightkeeper ashore is responsible for the good order and cleanliness of the precincts of the Shore Station. He is to ensure that the grass is cut regularly in the summer, that the refuse bins are emptied when necessary and that no refuse is allowed to accumulate. A careful watch is to be kept on the condition of the roofs, rones, down-pipes, etc.; and any defects or blockages which cannot be readily cleared are to be reported to Headquarters without delay.

(a) Lightkeepers ashore from Rock Stations may leave the Shore Station during their time ashore but should be at the Shore Station for the three days preceding the relief; absence extending beyond one night from the Shore Station should be notified Headquarters.

D.10. WATER TANKS.- The Principal Lightkeeper is to arrange with the Master of the relieving tender for the water tanks to be filled up before winter commences. So far as possible each water tank is to be thoroughly inspected outside and inside before the Superintendent's or Assistant Superintendent's visit and the state of the tank reported to him. This inspection is also to be reported in the Monthly Return.

D.11. MONTHLY LETTER.- The monthly letter from Rock Stations is to give particulars of the dates when the Rock was manned by less or more than three Lightkeepers on the Rock, e.g. when there was a Supernumerary Lightkeeper on the Rock in addition to the normal compliment or when a Lightkeeper had to leave the Rock in an emergency.

All of the Rock Stations I attended throughout my service were double manned three keeper stations. Three keepers were out on the Rock while three keepers were ashore. Shore Stations were houses built by the Board or on rare occasions, houses from local authority stock were rented by the Board but treated in the same way.

When we were ashore it became sufficient enough for the Principal Keeper ashore to notify Headquarters by phone if a keeper was going to leave the Station for more than three days. At most Shore Stations the keeper ashore would cut the grass empty the bins etc of his counterpart out on the Rock and likewise the Principal Keeper would do the same for his. We were obliging enough to do this for all if anyone was absent or ill at the Shore Station.

There was an additional payment for keepers serving at Rock Stations and this was called Rock allowance, my bank balance would be even further improved by this addition at the end of the month.

The notice from the Board had instructed me that I would be away from home for four weeks; I would need to make sure that I took ample toiletries, washing powder and personal items such as sweets tobacco etc for the period. Food would be provided the Board. Because of the remoteness of the Lighthouse, relief would be made by helicopter and as such, care should be taken to restrict personal items to those that could be safely carried as hand luggage. I was to fly from Glasgow airport to Kirkwall on Orkney and then to go from Kirkwall by bus to Stromness. Stay overnight in a B& B booked by the Board where I would be picked up by car to take me to the Lighthouse Pier where a helicopter would take me to Sule Skerry ...

©Scottish Lighthouse Museum

Chapter IV
Sule Skerry

For the second time Margaret and I said our goodbyes, the four days we had fleeting like a thought. There was time enough to do my laundry and get ready for my next trip away but time for little else. This time I would have to make sure that I took plenty of tobacco with me there were no shops nearby on this trip ... In those days both Margaret and I smoked, a habit we are ashamed of now. It consumed part of our income as well as our lung capacity but still the addiction demanded to be satisfied, at the time neither of us had enough motivation to drive the engine of will power to give up, the greatest shame being that we were inflicting our children to effects of passive smoking.

To get to Glasgow Airport, I had to catch a train to Paisley and then a bus would take me to the Airport, the bus journey at either end were the only parts that I would need to pay cash for, I would give the tickets to the Principal Keeper and he would refund me from Station Expenses.

I was to fly with British Airways to Kirkwall with a stop at Inverness; the aircraft was a turbo prop Britannia but comfortable and large enough for someone with little experience of flying on short regional flights.

The stopover at Inverness was for about 30 minutes; time enough to refuel and make additions and deductions from the passenger manifest, time enough also for me to see out of the departure lounge window two smaller aircraft .. I learned that they were called Trilanders and operated on routes to the remote islands ... taxi to the end of the runway then within a short distance become airborne; each had the three propellers, two on the wings and one central over the cockpit. This was the first time I had seen this type of plane and the first time I had seen an aircraft that needed such a small runway ... Its logical when you think of it though, some of the Islands would have had neither the space nor the capacity for larger aircraft but nevertheless much needed this airborne lifeline.

The aircraft was now re-fuelled and the passengers boarded for the remainder of the flight to Kirkwall.

I could have gone directly to Stromness from the airport; both the buses to Kirkwall and Stromness were waiting outside the small terminal building. I thought that this might be the only opportunity to see Kirkwall before I left Orkney. It was already mid-afternoon so I would not be able to see many of the sights that were illustrated in the tourist brochure, I had especially wanted to see Scapa Flow and the Standing Stones, but getting to either of them would have incurred a cost that I could not easily justify, even with our comparatively new found wealth. What's more and to my great dismay today was both a bank holiday and local holiday and most of the shops were closed. I did find a small cafe open and had a coffee and some ginger sponge, staying clear of the jam doughnuts that looked equally appealing.

An evening meal was to be laid on at the B&B so I did not want to spoil my appetite by eating too much. With very little open there was not much for me to see or do but I did hope that one day I would return and explore to the full what Orkney and the Islands had to offer.

I caught the next bus to Stromness, the holiday timetable meant that it was half past six by the time the bus arrived at the small port. I was given directions to the B&B and was happy to note that it was not far away and the incline up the cobbled street was not too steep. I rang the bell and a pleasant aroma of freshly baked bread mixed with a less distinguishable smell of what was sure to be this evenings' meal; wafted gently toward me as the door opened. An elderly lady beckoned me in. Her hair was near pure white but her gait and posture was that of a younger woman making guessing her age difficult, speaking with the soft tones that had a familiar lilt to it ... Bill Rosie also came from Orkney ... not making it any easier. I was shown to my bedroom and told that tea would be ready at seven. Also staying at the B&B were two Artificers from the Lighthouse Board; they would not be going out to Sule Skerry but to Stroma Lighthouse, it too was an island in the Pentland Firth but closer to the mainland. They would be going out to Sule Skerry at the end of the month, so I might meet them again there.

After tea I strolled down to the harbour and watched the evening ferry unload its mixture of passengers, cars and lorries. Further along the harbour I could make out a separate pier and jetty. Outside of a large shed like building I could see in the half-light the silhouettes of buoys of various sizes, some lay tilted and the largest of them standing upright and as large as the van parked just outside the gates, this undoubtedly must be the Lighthouse Depot I would have to report to in the morning. A minibus would pick us up from the B&B and take us along with the other keepers making reliefs to the lighthouses of the Pentland Firth.

With breakfast over it was not long before the minibus arrived, on board the minibus was John Kermode the Principal Lightkeeper of Sule Skerry and three other keepers going out to other Rock Stations, all were dressed in uniform. I could now see what the finished article would look like even if these did have a little wear to them. John's hat was different to that of the others; around the gold coloured lighthouse was a wreath instead of gold rope braiding. The hats were crowned with a white protective covering, the whole uniform similar to that of naval officers. On John's arm just above the cuff was a similar emblem to his hat and therefore denoting his rank

I introduced myself to John and apart from the acknowledgement that someone had actually spoken to him it became obvious that a full-blown conversation was not going to happen. It may have not helped matters and could possibly have caused deep offence when I referred to him as a fellow countryman; assuming his accent to come from the north of England when he in fact came from the Isle of Man, his correction to my mistake were his last words until we reached Sule Skerry. I learned a valuable lesson in those first few moments, never to try and make conversation when it is obvious that person is not interested or has other things on their mind and never to assume a person is from a particular area... I came across a few people who would take offence when they come from the next village and are associated wrongly. Thankfully both he and they are in the minority but it was to give me a strong indication of what the next four weeks would be like. It would be a severe test to my character and I hoped that I would be able to overcome any adversities that may be presented.

Before we could board the helicopter we had to sit through a health and safety video....I remember having to sit through a similar video while training in the Army, this time though we would be flying over the sea, so we had to pay particular attention to the evacuation procedure should we have to ditch ... Helicopters had just been introduced to the service for making Reliefs; over time as the service became more reliant on helicopters additional health and safety requirements would be introduced. We would be on the second trip, the two artificers going out to Stroma first.

Normally the Lightkeeper reliefs would take priority, but the helicopter had only one Keeper to pick up at Strathy Point near Thurso before going on to Stroma. The helicopter would refuel then take us to Sule Skerry.

We were not permitted to wander round the Depot, so I couldn't see much more than I could the previous evening; the buoys seemed even larger up close and in daylight had colour to them. The largest ones had cage like structures on the top and all had markers of some kind or another making them different from their neighbour. There were sounds of activity coming from the shed like structure but what was being constructed or repaired would remain a mystery. The quayside berth was empty and I was reliably informed by the depot charge hand assigned to helicopter duties that the" Pole Star" was at sea.

... The service had three Tenders, each had a designated area of service and had to supply the Lighthouses in their area with stores and supplies, they also have to service the buoys, minor less accessible lights and navigational aids. The "Pole Star" based here in Stromness covered the Northern Isles, Pentland Firth and sometimes Lewis and Harris. The "Fingal" based at Oban covered the Inner Hebrides to the Isle of Man and lastly the Flagship of the service The "Pharos" based in Leith covered the Firth of Forth to the Pentland Firth and south to St Abb's Head. She would also take the Commissioner's on their annual tour of inspection cruise...

You can more often hear a helicopter long before you see one and this occasion was no different. The sound of the Bolkow is as distinct as it is indescribable from the sound of other helicopters yet it would always be the welcoming herald on relief days. There would be a rush of activity as soon as the helicopter descended on its final approach to the helipad; those of us who were not on first relief were ushered back to the port-a-cabin out of the way. From the window I could see what was happening, I felt that I should watch the operation in order to be of use when the time came.

The red Bolkow helicopter landed without incident and the co-pilot exited from the front nearside door which had the livery emblem of the Lighthouse emblazoned on it; obviously this helicopter was designated for the Board if not owned by it, the Company logo Bond Helicopters on the fuselage precluding the latter. The charge hand joined him immediately and they exchanged a few words. Another Depot worker wheeled an empty four-wheeled trolley to the foot of the towpath and made his way briskly to the rear of the helicopter. He released a couple of clips and the rear doors swung open a little; he then secured the doors fully open using the stays that had dropped down. He removed a couple of items placing them on the trolley, and then took the trolley to the side of the port-a-cabin, returning to the helicopter with a fully laden trolley which he loaded into the helicopter. The two artificers climbed into the rear seats of the helicopter and buckled up their seat belts. The co-pilot checked to make sure all of the doors were secure and the aircraft was ready for takeoff before climbing in to his seat and closing his door.

The charge hand moved to the front of the helicopter to signal to the pilot that all was clear and the helicopter lifted and was away. This all happened in the space of four minutes and I assumed this achievement was gained by practice and repetition.

It would be another 30 minutes or so before the helicopter returned for us. Our turn would take a little longer because the helicopter would need refuelling. Two 45 gallon fuel drums were upended with a pump close by ready for the task. Aviation fuel, paraffin/kerosene has a distinct smell both before and after combustion, the traces of both were discernable even after the helicopter's departure. The Bolkow with its distinctive profile was affectionately, within the service given the nickname "The Paraffin Budgie".

The process of unloading and loading the helicopter was repeated for us; included in our load were four square, plastic red boxes; these contained the provisions for the rock for the next two weeks. I was happy to see my bag among those on the trolley.

I was bemused to note that the helicopter carried no Lightkeepers on its outward journey and now had one on its return. Apparently the Stroma Lightkeeper going out lived near Thurso but the returning Keeper lived in Stromness.

Once we were safely buckled into our seats it was not long before we were airborne. We were flying at about five hundred feet and soon the coast disappeared below us. It was quite a gray looking day with rain threatening at any moment; our destination was not visible. It would be another fifteen minutes before the Island with the familiar shape of a lighthouse at its centre would appear.

I would be going to one of the most remote Lighthouses in Britain, not that I was bothered about the isolation the only thing that did bother me was that as we got nearer it became clear that this flat outcrop of land about as large as a good sized field did not offer much in the way of walking. I would again have to find other things to occupy myself.

The nearer we came the more definition and colour the island took on, the green indicating that it was not quite devoid of vegetation even if there was nothing taller than grass. A windsock indicated that we were flying directly into the wind and we had to make only minor adjustments before settling on the pad. A loaded trolley similar to the one at Stromness was waiting, with two Keepers in attendance, both holding tight to their hats as the down draught threatened to relieve them. John was closest to the door so he would leave first. It was safer to use the one door, making it easier for the pilot to see to the safety of his passengers. I was quickly on his heels and without being asked I made my way to the rear to help with the unloading and loading.

I was also hoping that this show of enthusiasm might go toward repairing any damage caused by my faux pas of earlier. John did seem happy when he came to the back only to find he was not needed.

The helicopter was loaded and our stores and baggage safely secured on the trolley, there was no time at all to speak to the Keepers who we were relieving before they boarded and the helicopter was airborne once more. A sinking feeling suddenly gripped me as the helicopter lifted; it would be two weeks till the helicopter returned and four till it was our turn to be relieved. I was soon brought back down when John said irreverently; "Come on no time to dawdle, let's get this unloaded". I could see another keeper making his way towards the windsock, John acknowledging his presence with a wave.

We made our way up the path to the Lighthouse about a hundred yards away, the path had a dog leg to it and was not in direct line, other paths met at a junction and the widest path led to the cluster of buildings with the tower at the heart. This tower was a lot broader than Killantringan and looking up at the Lanternroom it seemed empty. The tower and the buildings lacked the sandstone finishing of the windows and facings. The only changes from white were the stone coloured Lightroom and the black of the astragals and dome.

At the base of the tower and gripping its girth was what I had been reliably informed was a quarterdeck. I could well imagine John pacing up and down on its roof and giving the command to repel all borders but other than that I failed to see the naval connection. The quarterdeck did not completely encompass the girth of the tower though it did have the look of originality to it as opposed to the adjoining buildings that seemed like after thoughts. In the background, the noise of an engine gave the building on the left the positive identification as being the engine room. There was no visible sign of a foghorn and the engine noise was more that of a tractor than the heavy thrum of the more substantial Kelvin fog signal engines, so it would not have taken Sherlock Holmes to work out that it was a generator of some kind.

The other clues that electric power was being used; were the outside lights installed at each corner of the quarterdeck and a mast some fifty yards from the tower supporting a cable which ran between the two. There was also light coming from two of the quarterdeck's windows. The pathway split either side of the quarterdeck and we took the path to the left, the passageway between the two buildings just wide enough to take the trolley.

We entered into a small courtyard at the rear of the quarterdeck; a door was open and the waft of cooking, relating that this must be the kitchen. Alistair MacDonald had just put the windsock away and came to help unload the trolley. John took his gear from the trolley and without word made his way toward a prefabricated building on the other side of the courtyard, it became apparent that it would be Alistair and I who would be putting the stores away. Alistair introduced himself and confirmed our status by saying; "I will show you where to put your gear, once we've put away the provisions."

Alistair took me outside, there was only my gear left on the trolley so I removed my bag and Alistair parked the trolley in the engine room. He showed me to a room in the prefabricated accommodation block.

"We have our rooms here," he said. This remark implied that we were not the only ones on the Island and sure enough and right on cue the sounds of labour could be heard. I could not imagine that there was anything so pressing as to have John start immediately to the task and the sounds certainly had the quality of being produced by more than one person. "We've got Artificers here for a fortnight installing equipment for monitoring the light, they sleep in the tower" Alistair went on, and finally stating that he would show me around after tea.

It became apparent that John had delegated Alistair to do my tutoring and that he was accustomed to this task, at this stage I was not in a position to criticise his actions nor was aware of all the facts to make reasonable judgment of him. In this I would give him the benefit of any doubt.

About eleven o'clock is break time throughout the service, so everything stops for tea, I helped Alistair make and serve the tea and coffee, John was now dressed in overalls and wore a blue engineers cap which gave him the appearance of Casey Jones and this would be his attire for the rest of the month with the exception of relief day.
True to form John said very little to anyone least of all me, so I did not feel singled out in any way.

Alistair and I had a major task to do before I could be shown round; I was taken to the watch room at the base of the tower. The watch room had wooden panelling and a desktop surface that followed the contours of the outer wall, underneath were cupboards all panelled with the same kind of wood. There were both VHF and HF radios on the desktop at the far end. A couple of shelves and some more cupboards lined the inner wall.

From one of the shelves Alistair took a hand held anemometer. On, the desktop beneath the window was an open book that was about the size of two A4's at the fold, the penned figures in the rows were for recording something that I was surely about to find out ...
Some Lighthouses are in ideal locations for recording and passing on meteorological information, Sule Skerry was one such Lighthouse and along with Muckle Flugga and Bell Rock featured as Weather Reporting Stations on the Shipping Forecasts. For the service of providing this information we were given a small reward, as it was not a duty required of us within the service it was not governed by regulation, because we were ideally suited in both our aptitude and our position we felt obliged to render our service Readings were taken every three hours by the keeper on watch and entered into the log, some calculations had to be made to work out some of the figures, then at the appropriate time these figures would be passed on to the collating station for further analysis before being sent on to the Met Office at Bracknell in Berkshire...
Alistair took from the table a notebook and a pencil and gave me the anemometer to hold. We left the watch room and went outside, just along the pathway next to the mast was a Stevenson Screen, I'd recognised this from my school days so I knew what we'd find inside. With this and the other equipment to hand Alistair did not have to make much explanation for the reason for our trip outside.

Before opening the screen he gestured for me to look at the sky and gauge how much of the sky was covered by cloud if I were to divide it into eight parts, the sky had not cleared since this morning so was still overcast, so this was simple 8 parts. "Now for the difficult part, what are the types of cloud and at what height are they?" he asked quizzically, noting the bemused expression on my face and the lack of an immediate answer he continued," don't worry there are books in the watch room that will help you with all the readings." He then made a few notes on the pad, he went on to say that it was difficult to be accurate on gauging the visibility the only land marks if any kind being a small island only just visible some eight or so miles away; it's only occupants the Gannets had whitened the top of it with their droppings otherwise it would have disappeared completely into the gray of the Firth. On clearer days I should be able to identify the coastline of the mainland to the South and extending westward towards Cape Wrath, looking eastwards lay the outlines of the Orkney's.

We opened the Screen and I was not disappointed to find the thermometers that I expected, what did surprise me was that there were four, two vertical and two horizontal. One of the vertical thermometers had gauze around its base. Alistair went on to explain that these were the wet and dry bulbs, noting the readings he said," the difference between the two readings after a bit of calculation will give you the dew point and relative humidity". I was equally bemused with this statement as I was when asked to identify the clouds, however I knew that Alistair would be quite capable of giving me instructions or there would be adequate tutorials in the watch room for me to be able to fulfil this task. Inside the screen was also a glass column that had numbers up the side this was for measuring rainfall that had accumulated in the rain gauge set in the ground about six feet from the Screen. Alistair checked the gauge even though it had not been raining since the last reading, he appeared satisfied that the bottle in the gauge lent support for his supposition. The last reading he made was to hold up the anemometer, the reading on the base gave both speed and direction.

Alistair suggested that this should be the last of the readings taken because; as it was a hand held meter the wind should be assessed; especially if it was gusting, before taking the reading, it would then reflect as accurately as possible the wind speed.

The page of his notebook was full and we made our way back to the watch room.

I had not noticed before but had just become aware that there was a lot of aerial activity; we humans were not the only inhabitants of the island. I had seen Puffins on wild life documentaries but never in life till now. The reason I did not see them earlier was because of the noise of the helicopter; it must have driven them to their burrows or kept them away from the island; their absence at the time of the relief a blessing to both the pilot and us. I did not have time at the moment but I knew I had to discover more about these enchanting little birds.

On our return to the watch room Alistair explained the significance of the log book and the recordings that went into it; Each row was separated into groups of five boxes, figures were entered into these except where they were not relevant A weather recording manual detailed what the figures represented with code numbers representing features that could not be presented numerically such as past and present weather, cloud types etc. We were not there to make forecasts but just to take the readings so we did not have to be educated in the world of meteorology. The hardest thing that we needed to master was the use of a slide rule to calculate the dew point and relative humidity; there was a table in the manual just for those calculations so I think the slide rule was there for the more academic Keepers to show off their skills. I mastered the skill before leaving Sule Skerry only never to use it again.

The log was now complete and the time was just approaching ten minutes to the hour, we had to contact the Collating shore station by radio and had to observe the silence periods just before and after the hour and half hour, so there was just enough time to pass the information. I was a regimental signaller in the Army so was used to using radio transmitters and the way to speak over the air. Alistair would be with me for the next set of readings at 1500 after that I would be on my own. I would be on watch this afternoon and the watches would follow the same pattern as Killantringan except there would be no days off.

Alistair still had a lot to show me but he was on kitchen duty so he would have to go back and finish preparing the dinner, "I hope you can cook, cause you're making the tea" he exclaimed. I would also be on kitchen duty for the rest of my month here...

I never had any formal training in a kitchen but coming from a family without a mother, we all had our chores to do. My sisters seemed to hold sway over the kitchen and my main duties there were to do the washing up, however my father liked to cook Sunday dinner and it was always something special like a roast of some kind. The girls were banished from the kitchen on the pretext of resting them or merely for my father to get peace from their interference. I was as always at the kitchen sink but he loved to show me what was needed and tried to involve my input as much as possible. He gave me the confidence to cook and even to experiment using a recipe book as a guide. So I was not too perturbed at the prospect of cooking for strangers...

Alistair had already taken the liberty of taking some pork chops out of the freezer for our tea; all I had to do was cook them. I asked Alistair if John preferred anything in particular to eat. His reply to that was, "If it's in the freezer, use it." It was a silly question really and I suppose I should have re-phrased it to mean was there anything in the freezer that was especially bought for the Artificers. His answer remained the same. He only added that the Artificers expected to have a cooked breakfast so who ever was on kitchen duty would have to make sure that bacon and sausages were taken out to thaw for breakfast in the mornings as well as making sure that something was taken out for dinner. There was no fresh milk so powdered milk had to be made up in a jug; the art was not to make it too thin or too thick.

The keepers would normally have cereal or porridge and toast but have a cooked breakfast on a Sunday ... This was the routine that I found at all the Rock Stations that I served at. We always made sure that we were well provisioned and ate well, there were at least two meals a day that had meat and at some stations especially on the evening after the relief we had fresh steaks for tea. The only time I can ever remember being fed as well was when I was in the Army. Thankfully there were no vegans or vegetarians because I'm sure they would have become undernourished during their stay. Any piece of fertile soil on the Island was otherwise put to use by our fellow inhabitants and so useless for the production of vegetables.

Alistair and I completed the dinner routine by doing the washing up and making sure that the kitchen was tidy. After, he gave me a tour of the Tower starting with the basement. The Lighthouse was undergoing the transition to becoming automatic, the new light had been installed and the sensor system was in the process of being tested. The lamp itself was powered by a form of Acetylene gas that was stored in bottles mounted around the walls of the basement. Pipe work fed the gas up to the lamp. We climbed the stairs to the first floor: for indeed this lighthouse had floors or storeys hence the reason for the towers broad girth; on this floor was a sitting room similar but smaller to the one in the quarterdeck. It was finished out in wood panelling and had a couple of cosy chairs on the inner wall. Underneath the window was a small table. Rows of books lined the tiers of shelving above the chairs, quite a small library I thought being sure to need to visit here regularly throughout my stay. Judging by some of the titles, I would not be disappointed in my search.

On the next floor was a bedroom, again it was panelled, but this time a set of triple bunks occupied the inner wall. A wooden ladder allowing access to the upper bunks was secured to a panelled end wall that had a storage cupboard or wardrobe set into it. Curtains draped down over each of the bunks that allowed the occupants some semblance of privacy. I was glad that we had other accommodation ... Up until recently the keepers would have slept in the tower but because of the amount of work that was going to take place during the Automation, it was felt that the Keepers would not have enjoyed uninterrupted sleep during the day and therefore it could possibly affect their efficiency to perform their duties. Even though at this present moment our role was reduced to just monitoring the light, we did have to occasionally put the light on and put the light off during the time that the sensor equipment was being calibrated.

Moving up to the next floor, it was a repeat of the last. So the tower could have accommodated six people. I understood from reading the regulations that these types of Rock Stations had three Keepers at a time residing there, the other three bunks I presumed must be for visiting Artificers etc. Alistair could judge by my expression that I was unfamiliar with this kind of set up and went on to say that if I thought things were a little cramped here I should see The Bell Rock or Skerryvore; at least here I could walk outside. They are Pillar Rocks and only the Tower can be seen except at low water.

I consider myself very fortunate that I never got to see or serve in either of those Lighthouses; in some keeper's eyes that may have made me a lesser Keeper but that was just service ribaldry. They along with Stations like Muckle Flugga, The Flannan's and Dubh Artach were the most notorious throughout the service. The Flannan's gained its notoriety for losing its Keepers during a storm. In December 1900 a year after it was first lit The Hesperus was carrying out the relief, The Assistant making the relief discovered that the Station was empty. No trace of the missing Keepers was ever found. It still remains a mystery but several theories prevail amongst us Keepers; the most popular was that the Flannan's was bad for exposure to storms. The Atlantic swells had nothing to break their force before hitting the Islands so giving rise to freak waves that can sweep up and over the rock faces at some considerable height above sea level. If you have ever heard tell or have experienced what is known as the seventh wave phenomenon where it supposed that every seventh wave has a force larger than that of its predecessors, and you magnify that by a factor of fifty it goes somewhere to explaining how high and with what force these waves have.

I can't imagine and there was no evidence to support the theory that a wave swept to the level of the Lighthouse itself and swept the Keepers away but it is possible that the Keepers were attending to equipment at the landing or even at the stages above when they were swept away.
The Flanna's were made automatic before I joined the service but I did serve with Keepers who had served there. Each of them emphasised how stormy the place could get and how high the waves could reach and all of them said that the only way to believe it would be to stay there. Dubh Artach was notorious in the service for being the Lighthouse that was hardest to relieve, many a Keeper had to endure overdue reliefs because of bad weather, and this was another Lighthouse that was thankfully made automatic before I joined the service...
Above the second bedroom and now used mainly as a storeroom was what was once the kitchen. It seemed strange to have the kitchen above the bedrooms but it made sense once you thought about it. The greatest possibility to the risk of fire occurs in the kitchen; heat, fumes and fire have the tendency to travel upwards, anybody trapped above a fire in the tower would probably be engulfed by it with little chance of escape even the balcony would not offer much respite as the heat would be transmitted through the stone. Lighthouse Keepers had to evacuate from the Chicken Rock Lighthouse on the Isle of Man because of fire that started in the kitchen, the Principal Keeper had to abseil from the balcony after the heat became intolerable.

Above the now storeroom was the Lightroom. It differed little in form from that of Killantringan. It was what was not there rather than what was, that made the difference. This room was almost twice the size and empty in comparison, you could still make out where the winding mechanism once stood, only a small column replacing it and going up to the light itself. An electrical switch box with some instrumentation of some kind was fixed to the wall and had cables leading to the column and to the grating above on which was housed a stubby little light like a miniature version of the top of the tower we were now in. The cavernous space of the Lanternroom made small of both that light and the light that occupied the centre. Where the lenses and their carrier at Killantringan occupied nearly the whole of the Lanternroom; here the whole mounting was not much taller than a man, the pedestal came to about chest height and the lenses the rest; less than three foot across and about two feet high, back to back and hinged in the centre. In the centre space was the light source itself, not electric but a gas mantle. Three mantles were set on a rotating spring-loaded mount that had a balsa wood fuse trigger. When the mantle was damaged; the flame of the light would burn through the fuse triggering the mount to rotate to the next mantle thus keeping the light. The lenses did the same job of magnifying the light as those at Killantringan. There was no blanked out section of the lantern here and likewise there was no need for curtains or blinds.

The system in use here was known as Dalen, as the acetylene went up to the mantle it powered by pressure the lens carriage. Making the carriage bearing free and floating it on a mercury bath reduced friction.

So this was to be the shape of things to come. I was aware that there was an automation program but at that time automation of the whole service seemed very distant. The system in use here was still in its infancy and nowhere near to being monitored from the proposed remote monitoring centre in Edinburgh. I would definitely be reading a few books, as there seemed very little else to do. There was little for me to learn in the way of Lightkeeping here either, it would be more of a case of tolerating the potential for boredom and with my staying in the kitchen throughout; there would also be the monotony of routine to contend with.

Alistair would help me with the next weather report at three pm; in the meantime he would help John and the Artificers. From what I could see from an earlier encounter that day; it would mainly consist of standing watching them and giving the occasional supporting comment or on the very odd occasion going to fetch them a tool or piece of equipment. I on the other hand would spend as long as I could observing the puffins.

Once you left the confines of the Lighthouse and its pathways the ground was pockmarked with holes, most of them only just visible beneath the tufted grass that was only as high as the wind and the salt spray would allow, but lush just the same. Coming from these burrows was a faint sound that I can only describe as being like a chorus of MP's sounding off approval at Prime Ministers Question Time.

I had not seen any birds entering or leaving these burrows as yet and I hoped that I would be able to find somewhere to conceal myself in order to watch them. I decided the best place might be at the side of the engine room, figuring that the noise of the engine would conceal any noise that I made and the warmth from the exhaust that protruded from the wall having some of its heat transmitted down the wall to where I had found a reasonably comfortable place to sit.

So far I had only seen them flying at a distance. In flight it seemed as though they were constantly struggling to remain airborne. Their short stubby wings flapping inherently fast as with other members of the auk family, like guillemots and razorbills. They also have the same black and white colouring. Even at a distance the distinctive beak was noticeable even if its coloration not quite so. I was about to adjust my position when a bird flew towards me and landed some twenty feet or so away, in its beak I could see some fish. They appeared to protrude from both the left and right sides. My presence was ignored assuming it had been detected and the bird continued on its way to what I assumed was its burrow. It appeared to waddle quite fast with its head down and then it vanished into the burrow. It must have been the right one because there was neither sight nor sound of commotion or eviction... I often wonder whether it's just us humans that have the tendency to forget where we live and at times stumble into the wrong houses. Not of course that it has ever happened to me, as I always extol the virtues of sobriety, but having seen and heard of others of a lesser persuasion doing just that. My imagination could run riot just thinking about it, e.g. Maybe it was man that caused the dodo to become extinct, not by eating it but by feeding it rum; feeling sorry for the poor creature with its comical shape and posture. Its death caused by wandering drunkenly into the wrong nest and being pecked to death by an irate neighbour or wandering into the right nest only to be henpecked by its frustrated mate ...

I would spend what spare time I had split between reading and watching the antics of the birds. Surveying the Island was going to prove difficult and restricted to wandering the pathways to and from the station. To wander off these paths was not only dangerous and disturbing for the birds but also for the wanderer. A foot carelessly placed, could quite easily cave in a burrow killing the occupants and or twisting and breaking an ankle. Although the puffins appeared comical and endearing they were here for the purpose of bearing and rearing young and at this time they would be at their most vulnerable. The whole purpose of excavating burrows was to protect themselves and their offspring from predators. I had witnessed blackback gulls mob a returning puffin in order to steal its catch. At a distance I had seen a skua; the puffin's most dangerous predator do the same. This would be the first and last time that I saw a skua. They are rarely seen south of the Northern Isles and the Outer Hebrides at first glance they look like juvenile black backs only larger. The skua seemed content just to deprive the puffin of its catch, however I had been told that when the skua has young to feed the situation would change and the skua would take the puffins themselves. It was not unknown for a skua to knock a puffin out of the sky, but the simpler method was to patrol the air above the burrows and wait for an approaching puffin laden with its catch to return to its burrow. So that was the reason for the puffin to have its head down and make a dash for the burrow.

I would not be on the island when the young would fledge and the only time the young could be seen was when it was pitch black, even then under torchlight it was hard to tell adult and chick apart. Taking weather readings at night was a quite hazardous job and impossible without the aid of a torch. Not just for seeing the readings but to avoid the puffins that appeared to be everywhere. Their calls were louder now without the muffling effect of the burrows. You would expect them to scurry away at every footfall but that was not the case; they would only move when the light shone in their direction. By day they would have to avoid the attentions of the blackback's and the skua's by night the feet of the Keepers.

I was drawn to the conclusion that these birds were so full of purpose that little distracted them from their path. They would unerringly fly the gauntlet of blackback's and skua's to take the most direct route back to their burrows. All in all it seems a desperate life for these plucky little birds but so long as the numbers of blackback's and skua's did not increase then their burrowed existence should ensure that enough would survive to continue the process. When the lighthouse was eventually de-manned there would be fewer disturbances for them but no one to witness their struggle.

Watch Keeping on Sule Skerry was not very demanding, with no fog watches and no machinery to wind up, the only tasks required of us were to make sure that the light was both shining and rotating, something you did not even have to go upstairs for. You could look out of the watch room window and see the reflection of the light as it passed and climbing the stairs would only risk disturbing the Artificers from their slumber.

Apart from the weather readings every three hours, at the end of each watch we would have to check the generator for oil and pump fuel into the header tank. There were two generators only one of which was running at any one time. Their use was rotated every Saturday when the engine that had been running all week was given a filter clean and an oil change , the commutator and other parts were also cleaned.
Kitchen duties did not involve just the cooking; I had to do the housework as well. The quarterdeck living room, tower living room, hallways, toilets and the kitchen all were cleaned on a daily basis. The water for washing and toileting was collected from rainwater and never in short supply its only drawback was it was harsh and hard to get the lather from.
Drinking water had to be monitored and restricted, it had to be delivered to the island in carboys; each container held about 20 gallons of water. Some empties were stored in the engine room but they were of discontinued use. The Pole Star made a delivery of water every six weeks or so depending on the demand made by extra personnel. The water was decanted into a separate tank and the empty containers returned to the ship. We were expecting the Pole Star to make a delivery of water and other stores and equipment within the next two weeks. It would always be weather permitting and at this time, in this particular year the weather would prove both unseasonable as well as unpredictable.
The week that the Pole Star was scheduled to arrive started with a snowstorm followed later in the week by gale force winds. If the trip had been scheduled for the week previous then it would have been made during a few days of glorious sunshine. As it was, the storm from the west had increased the swell that no attempt could be made till the following week. Although the wind speed had lessened the swell would take longer to dissipate and the ship's captain Neil Morrison would call on Tuesday evening to see what the landing was like. I was on watch that evening but it was John who took the call. As I would be on watch in the morning when the final decision on a landing would be made; John took me down to the South Landing to check its condition and to show me what to look for when I would make my report to the Pole Star.
The South Landing was a gully that was either man made or natural; time had worn away traces of human endeavour if that were the case; the surface of the black stone uniform in appearance. A pier had been built which had steps leading down to the water. On the landing was a derrick painted in the red oxide primer; it was used to load and unload the ship's tenders. A landing was faced with one of two problems; there would either be too much or too little water in the gully, too little water was easier to solve by waiting for the tide. Too much water in the form of swell could drive the tenders onto the rocks at the end of the shallow gully. Metal rings were secured into the rocks around the gully for tying the boats mooring lines to; they would help stabilize the small boat during off loading, but they would not be enough to compensate for a heavy swell. If the water were at a point marked by one of these rings at the end of the gully; then it would be too hazardous for a landing. At the present moment it was just at

that point. Should the case be the same in the morning then it seemed unlikely that there would be a landing.

I was making the Artificers their light supper of cheese on toast when John received the call from the Pole Star so I was not in a position to hear the conversation. John only relayed to me that the Pole Star would call again just after seven am and that I should pass to them the wind speed, direction and the state of the gully. I was to call him if there was any problem or should the Master wish to speak to him.

It felt as if the responsibility for a safe landing was in my hands and it was with this though in mind that I turned in somewhat worriedly at the end of my watch. It still worried me when I resumed watch in the morning. The weather readings had already been taken and passed so all I had to do was to go down to the landing and see how the gully was.

The swell had lessened to the degree that it was four feet from the ring; the wind was blowing steadily at sixteen knots from the SSW. To my mind it seemed as this would be a borderline case, and I hoped beyond hope that my own judgment would not be called upon. With these thoughts still in my mind it was a nervous Keeper that awaited the voice to come over the radio.

It came at six minutes past seven; "Helloooo Sule Skerry, Sule Skerry, Sule Skerry."

"This is the Pole Star calling Sule Skerry." "Is there anyone awake there over?" At that moment and hearing those words with that intonation all my worries seemed to evaporate. Whereas I had expected to see Bill Rosie looking like a character from the Vital Spark, here was a voice worthy of Dan McPhail coming to me over the airwaves. It took me a moment to compose myself before answering. "Roger Pole Star this is Sule Skerry." "Go ahead over." I knew then that he could ask anything of me and I would not be so hesitant to give answer. I passed over the weather and landing information only hesitating on where the water was in the gully. It did not take him long to come back with. "There will not be a landing there today," "I will call again in two days time. Pole Star over and away."

The decision for a landing was always going to be his and his alone, so there was never anything for me to fret over. All I need do was to pass accurate information across. His was definitely the right one as the wind picked up throughout the day and the storm that followed would last till just before our relief, so I never got to see the Pole Star and its Legendary Captain; Neil Morrison.

The routine of cleaning and the harsh water had taken its toll on my hands; the skin between the fingers had hardened and cracks had formed. I was now at risk of contracting dermatitis if I had not already done so. I had hoped that John would have mellowed over the past week or two but was disappointed to find he was worse. I asked him if there was any barrier cream at the station and his reply was to ask me "How old are you?" I had an idea where this was leading but played along anyway "What has that got to do with it?" I replied. He repeated the question; this time I gave him the answer he wanted "twenty six". "What does a grown man like you want with Barrier cream?" At that point I showed him my hands. "Do you recognise that?" showing him my hands. I went on to say to him that it was not his respect that I was looking for but civility owed by one human being to another. I pointed out to him that although the Board expected to receive a report from him about my conduct and suitability, I too would have to submit a report especially when I would be prevented from doing my duty because of an avoidable medical condition and have evidence to support the fact.

To this day I am not sure what possessed John to behave in this manner, again I will give him the benefit if the doubt by supposing the Board had directed him to act in such a way as to apply stress so that it might uncover any failings in my character. This I could understand but if so, on this particular occasion he overplayed the part. He did mellow somewhat over the last week even to the point of holding conversation. But the heavens were to present the final verdict on the morning of our relief...

We had just had breakfast and John and I were the Keepers being relieved, so we had little to do but get ourselves ready for the relief. The two Artificers that I had met in Stromness did not come out to the Island; the equipment they needed was still aboard the Pole Star. There were just the three Keepers on the Station and two were due to leave in an hour or so.

The day was going to be fine and the sun had the warmth typical of late May, a few days ago I had spent my birthday out on this lonely outcrop and it passed without notice but I liked it that way especially as I had the promise of Margaret making up for it when I got home.

John and I had prepared the trolley and taken it with our baggage and empty provision boxes down to the helipad. John was wearing a white shirt with his sleeves rolled up. Our spirits were lifted by the prospect of the relief and we began a conversation.

Retribution comes in many forms and on this occasion it was not directly asked for but while we were talking and John was relaxing; resting one leg on the platform of the trolley; a bird unloaded the largest amount of droppings I'd ever seen from one bird; directly onto his uncovered and balding pate. The overspill stained the shoulder of his once white shirt.

The restraint I had to show in not bellowing with laughter was phenomenal, as it was I could not help let a smile broaden my face. Most surprising of all was that he spoke not one word but let out a smile in return.

I had to contact the Board to find out what where I would be going next, John while still in his belligerent mood had said I was responsible for finding out for myself what would be happening to me after the relief. This was not true of course; when I contacted the Board they were surprised it was me making the request and not John, however they did tell me that I would be having a week or two at home before going to my next assignment. A decision had not been made as yet but they would let me know in due course, probably while I was at home, in the meantime I should enjoy my time at home and make full use of it.

This would always be the case for Supernumerary's they would have to relieve Keepers who went on holiday or in an emergency had to be taken off because of illness.

In the quiet periods I might find myself ashore for weeks at a time and then suddenly have to go because of an emergency...

At this moment waiting for the sound of the "Paraffin Budgie" the prospect of a week at home was more than enough to lift my spirits, paling to insignificance the test of endurance of the past few weeks.

Chapter V
Ruvaal

However enjoyable the few weeks break, I still eagerly awaited my new assignment. If by going to as many Lighthouses as possible in as short a time as possible brought me closer to the day that I would be appointed, then so be it.

I was to go to another Rock Station only this time it would be on the Island of Islay, I would have to fly from Glasgow to Islay, and stay overnight in Bowmore and then get a taxi to a Distillery called Coila. There a fishing boat would take me up the Sound of Islay to Ruvaal Lighthouse. I would be relieving a Keeper who had taken ill. I would only be going out to Ruvaal for two weeks this time.

The flight to Islay was only twenty minutes; barely time to get comfortable if one could get comfortable in this small aircraft. It seemed as though this airline operated aircraft according to booking and demand. I had seen larger aircraft operated by the same company just a few gates down from our departure gate. Even on this twenty six seat aircraft there were empty seats, so demand for this afternoon's flight was low.

We flew over the bottom end of the runway at Islay on our approach to land, making three right turns before lining up with the runway. The visibility was clear this afternoon thank heavens; I had overheard a conversation between two fellow passengers one stating that Islay had its problems for landing aircraft. Apparently the hills there have the habit of forming mist or fog that hampers safe landing. A couple of years after this first my first flight to Islay, an aircraft flown by the same company crashed on its approach and there were fatalities. After that incident if the pilots could not see the runway on their approach over the beacon at the bottom then they would abort the landing and fly back to Glasgow. Today, we made a perfect landing.

I stayed in a B&B just outside Bowmore, my room at the front offered a magnificent view over Loch Indaal. Across the Loch I could see the white painted walls of a Distillery. A mile or so up was a similar Distillery only this time it was among houses painted the same snow white. In between the two on what seemed like a small promontory there was a stubby lighthouse yes you've guessed it also painted white. As the taxi that brought me from the airport to the B& B passed through Bowmore I noticed that about 80% of the houses were painted white. I could imagine that whoever sold the paint was onto a good thing, I also reckoned that because this part of the Island was open to the west with only low hills or no hills to protect it from the weather then he would have repeat and regular business.

I did wonder briefly if the lighthouse I was looking at was the one I would be going to but a map on the inside cover of a tourist brochure that I'd found in my room soon divulged that it was Loch Indaal Lighthouse. On the map I could also see that the Distillery across the Loch was Bruichladdich and the Distillery in the village across the way was Port Charlotte, there were a further six more Distilleries on Islay and one on the neighbouring Island of Jura each producing Single Malt Whisky that were also singular in character.

I am not a teetotaller and I beg forgiveness if any earlier comments may have led you to believe otherwise, but Whisky has never a drink that I have been overly fond; I'd much prefer a Brandy and Lemonade. An expensive round you may say and you'd be right but I had acquired my taste for this tipple while I was stationed in Germany. All spirits in the NAAFI were the same price as the beer, it was only the mixer that bumped up the price and even at a Mark for a glass it was still cheap. Marriage and common sense put paid to drinking anything more than the odd rare glass though I still find nothing more refreshing than a glass of Ice Cold lager on a hot day so you may during the height of summer find one or two in my fridge just for that rare occasion.

An ordinance survey map would have shown me where the Lighthouse was so in the absence of one I could only surmise its location. I could see Coila marked on the map only because there was a Distillery there and apparently a road passed through it on its way to yet another Distillery at Bunnahabhainn. It was somewhere beyond the Distillery that I believed Ruvaal to be.

A small wooden fishing trawler was tied up at the small pier at Coila just as foretold in the letter, a man about my own age introduced himself as being the skipper, on board a young lad was busying himself about the deck. A few lobster creels were stacked just forward of the wheelhouse but there was nothing to suggest that they would be pursuing any serious fishing after dropping me off at Ruvaal. The skipper was busy in the wheelhouse and the boy still busied himself, possibly in an effort to avoid talking to this Sassenach. I had overheard them in conversation speaking in what I thought must be Gaelic and the word Sassenach did enter into it, they seemed pleasant and hospitable enough so I did not feel it was rude of them to speak in their native tongue. - At a later date I did find out that Sassenach was a term used offensively to refer to Englishmen; however I was also told that it is also a term used to describe non-Gaelic speaking people especially those who live south of the Highlands. With my accent and my lack of the tongue I fell into both of those categories, my ignorance did not allow me to take offence then and my respect would not allow me to take offence now.

The boat chugged its way up the Sound, as we passed Bunnahabhainn the young lad chirped up, "We're going passed the last outpost of civilisation now."

Had I not just served at the most remote Lighthouse in Britain, I might have shown more interest or appreciation of the statement as it was he merely concluded by saying. "The road ends there, after that there is not even a track to take you to the Lighthouse." He was now pointing to the finger of the tower sticking up from a headland some five miles further up the Sound.

As we approached the landing I could see a tall bearded figure dressed in uniform but hatless; give a wave in greeting. The smile on his face and his bearing a sign of affability was in stark contrast to my reception in Stromness. This man looked even younger than either Bill or John so I had doubts that this was James Budge the Principal Keeper; even further doubts were cast when an older man came down the slope from the Lighthouse. With the absence of any other craft and the landing not big enough to take a vessel even as small as this fishing boat combined with the actions of the lad as he began releasing the clamps that secured a flimsy inflatable to the front of the wheelhouse, I seriously began to wonder if my future as a Lightkeeper would come to an abrupt halt within the next few minutes. It's not that I have a fear of water but just the respect one has when they have not learned how to stay afloat let alone swim.

Just to add to my growing discomfort; it would be this young lad who would paddle me in this raft that I thought barely big enough to support the weight of one person let alone two across the thirty or so feet from boat to landing.

Surprisingly it was quite stable once you managed to clamber into it and thankfully the skipper held the inflatable close to the side during the process so there was only one moment when I thought the action might tip both the lad and me into the Sound.

The bearded gentle man was indeed Jim or Jimmy Budge, the arm of his jacket bearing the emblem of rank establishing this beyond doubt as it was thrust supportively in my direction. Once again soft Orcadian tones bade me welcome, "Hi, you must be Peter." "Glad to see you dry and in one piece," he said and at the same glancing across at the lad and giving him a ribbing wink and a smile. The boy smiled in return but said nothing. "Will it be lobster for tea tonight Iain?" he shouted to the skipper. "Aye, but

I'll no be back here, we're off to check the creels at McArthurs' Head" he replied. McArthurs' Head was another Lighthouse at the other end of the Sound, an eight to ten knot current swept down the sound carrying plenty material for the lobster to feed on and the current lessened there as the sound opened up.

"No matter," Jim retorted, "we'll just put up with our T Bone Steaks then," adding "Have you time to take some breakfast with us Iain." I did not believe for one moment that the offer of breakfast was genuine as it must have been past ten o'clock, however I think the hospitality was real enough. "No thanks Jim, wee Davy has to go to Port Ellen this afternoon," Iain replied gesturing towards the lad in the raft as he came alongside and grasping the line thrown at him in the next moment.

I thanked them for bringing me safely ashore and we bade them farewell, Jim introduced me to another Iain the Local Assistant. He lived in Bowmore and although he spoke in soft tones his accent differed greatly from Jim's but not so great from that of Neil Morrison, I think the link that bonded the two was the common language.

From the landing to the Lighthouse up the gentle concrete slope was less than a hundred yards, reaching the top of the slope we in unison but unrehearsed, glanced back towards Iain and Davy who had already turned the fishing boat and headed back up the Sound, giving them a final wave we entered the small courtyard of the lighthouse.

There were similarities between Ruvaal and Killantringan but the differences were greater. The greatest difference was the height and narrowness of the tower itself, it appeared taller than Killantringan by some twenty-five feet or so but much narrower; liken to compare an index finger to a thumb. It was completely detached from the other buildings and separated from them by a path. Once again a boundary wall encompassed the whole station; painted the familiar snow white that contrasted so well with the green of the manicured lawns and the darker green of the doors. Absent was any sign of fog horn or receivers; the only red was a water tank mounted on a stand and what I took to be a fuel tank with a pipe coming from the top that crossed a pathway and fed into the side of the engine room. The thrum an engine supporting the fact, this engine had a much heavier pitch to it, not as heavy as the Kelvin's' that powered the fog horn at Killantringan but heavier than the Listers' that generated the power at Sule Skerry. What was obvious at a glance that there were fewer buildings here than at either Killantringan or Sule Skerry and the Light appeared to be the only purpose for the Keepers presence here. At the moment though and as was the norm I would be shown to my room given chance to unpack before meeting up again with Jim and Iain for coffee.

All the Keepers had their own bedrooms at this station with a couple of bedrooms to spare and it was to one of the spare bedrooms that Jim led me to. The warmth and affability shown already by Jim and the fact that I would only be here for two weeks were prophecy to my passing a pleasant time here at Ruvaal. Iain was to help make this even better by staying in the kitchen for the whole time. It was entirely his preference and choice no matter how much I tried to persuade him otherwise. Jim and I would do the remainder of the cleaning at the Station, what little there was seemed to have the pristine and well maintained look about it. Iain took umbrage when I thought of cutting the grass; that was his pride and domain even though he had enough to do in the kitchen and he would spend his spare time in the afternoons grooming his Kingdom in one way or another or lie on a beach blanket and soak up the sun.

With coffee break over and with just as little haste Jim took me a tour of the Station starting with the Tower. The stairwell seemed steeper but that I assumed was due to the fact that the stairwell was enclosed rather likes a turret of a castle. Although light came through the windows that staggered their way up to the Lightroom in the same way as Killantringan, there was not enough to not need lighting even in daytime. It was quite a climb in comparison and as we neared the top Jim noting that we both showed the signs of exertion said, "thank God we don't have to wind up here." I did kind of ponder the same thing but having come from Sule Skerry where they also had no machinery to wind I was quick to conclude that any motive power was more likely to be for electrical use rather than mechanical.

The Lightroom was about half the size of Killantringans' with barely enough room to fit the ladder to the Lanternroom the electrical switchgear and a small table. A door led out onto a narrow balcony that Jim insisted on taking me out of before progressing any further. I think he was trying to find out if I was afraid of heights, like most people it's not the height itself but the fear of hitting the ground from such a height that bothers me. My first reaction was to look over the balcony to the ground below then I turned to look up at the astragals of the Lanternroom and to the dome beyond. Suitably impressed that I had seemed to retain my composure Jim guided me over the vista that opened up before us as we traversed around the balcony.

Before us across the Sound of Islay was the island of Jura and from this far up the Sound the two peaks known as the Pap's of Jura appeared a little deformed, the left breast larger and more voluptuous than the right. Down the Sound and at about five miles distance the Distillery at Bunnahabhainn was our nearest neighbour. There was little to see on the landward side, a hill obscuring anything beyond. A line of telegraph poles wended their way back from the Station in the general direction of the Distillery this was our main source of communication. At about fifteen miles to the north from the mouth of the Sound lay the Island of Colonsay to the Northeast and beyond Jura the faint outlines of the mainland could just be seen. Jim told me that between Jura and the mainland was the gulf of Corryvrechen, a notorious stretch of water that was feared and respected by mariners because of its contrary undercurrents, tides and whirlpools.

I was told to look out for the Lights of Dubh Artach and McArthurs' Head Lighthouses that should be logged at midnight in the Returns book. Had we brought either a telescope or a pair of binoculars out onto the balcony with us we would have been able to see Dubh Artach just to the left of Colonsay Jim assured. Whether it was imagination, anticipation or just good eyesight but I thought I could actually see very faintly something that might just be Dubh Artach just where Jim suggested I look. I did not wish to appear a braggart so I kept it to myself, after all it might just as well have been a yacht, and I did however want to investigate and confirm my thoughts later; with the aid of either telescope or binoculars that I now knew were at the Station and had full permission to use.

We made our way back into the Lightroom. Jim's instruction on how to put the light on and off took all of two minutes, its duration dictated by the soft Orcadian drawl and Jim's leisurely approach to life. Whatever Jim's demeanour the Station was immaculate and the smell of freshly painted surfaces still purveyed the air weeks after the task had been completed.

We climbed the ladder to the Lanternroom; instead of net curtains there were blinds that were pulled down to shield the sun. The lenses and its carriage were smaller than Killantringans' but larger than the Dalen at Sule Skerry. The lenses alternated between red and clear, thus giving the light its character of three white flashes and red flashes every fifteen seconds. One section of the lenses was hinged and opened up to reveal the light. A light bulb the little larger than a football sat on a pedestal, its filaments more numerous and pronounced than the domestic version otherwise similar in appearance.

Looking at the astragals I could see that a few of them had been used as target practice by the local bird population and knew instinctively that there was an alternative reason for our sojourn out onto the balcony. Noting that my attention had been drawn to the marks on the glass, Jim ventured, "That's a big problem at this time of the year, and one that we'll have to deal with on a regular basis but not today." This time he had to be assured that my earlier reactions on the balcony were not just a front, he asked bluntly "Would you have any problems with working at heights?" Pausing a moment and giving careful consideration to my response, I replied "No, not really", the not really part of the answer indicated to Jim that I did have some reservations, he was quick witted enough to figure out my concerns and responded, "Don't worry we have to wear full safety harnesses these days" and "I'll make sure you keep safe while we're out there." Reassured by this, I reaffirmed a little more convincingly that there would be no problem.

Before leaving the Lantern and Lightrooms we took cleaning rags from a cupboard below the ladder and wiped the lenses and surfaces, the whole task taking about fifteen to twenty minutes.

Going down the tower was easier but more hazardous than the climb, the danger lying in that momentum could overtake stability which would then be hard to stop. .I foolishly experimented the experience only once during my time at Ruvaal, crashing headlong into the cupboard behind the door at the entrance, thankful that my exploits were shielded behind the closed door, had it been opened I'm sure that I would have been pitched out of the entrance to land sprawling on the lawn in front of the living room and kitchen windows...

Our next port of call was the engine room; this time only two engines about half the size of the Kelvin's occupied the floor. The one on the right providing the power while the other quiet, the silencers of the exhausts dampening the sound of the engine down to the level of that of an automobile. As usual our job was to check the oil and water and with these engines the oil and the engines were changed fortnightly just before the reliefs. There was no brass work on these engines so cleaning them took very little time or effort. This station was on the schedule to be automated within the next two years, the timing would depend on bringing power over the hills from Bunnahabhainn.

The only thing I needed to be aware of was that the engine room had a fire safety system that was effective but hazardous to man. In the event of fire in the engine room after detection an alarm would sound giving anyone two minutes to evacuate the engine room before halon gas flooded into the engine room expelling all the oxygen and extinguishing the fire. The alarm system was checked weekly and recorded in the Returns book.

That was all there was to this Station so it appeared as though I would have plenty of spare time on my hands, I would try and explore as much of the local area that I could without breaking the regulations or endangering myself, although on first appearances there seemed little in the topography that could prove hazardous. I had no intentions of swimming in the Sound so there was no risk of me getting carried away by the current. There were no cliffs to negotiate, quite the contrary; from the balcony, I could see that there were dunes and beaches from the Station to the head of the Sound and beyond the Sound the northern shore seemed like a plateau of flattened rock that tapered out then stopped abruptly before reaching the head of the Sound, its other end obscured by the hill that traversed from the shore all the way back towards the distillery. I was hopeful of the prospect of some rewarding beach combing and eagerly anticipated my first expedition, I would check out the Station this afternoon and make myself familiar within its confines and leave venturing further afield till tomorrow.

Next morning as envisioned, Jim and I gathered our cleaning materials and our harnesses in preparation for tackling the astragals. As there was no usable water in the tower we had to fill our buckets in the kitchen ... Rainwater that collected in gutters at the base of the dome was piped into three tanks on the balcony but these tanks inside's had become encrusted over time with Verdi Gris so the cleanliness for the purpose could not be guaranteed ... We also had to put on our harnesses at ground level as there was more space down here to do so. Jim showed me how to put on the harness and checked that it was secure before we made our way up to the balcony, being careful not to spill any water adding to our tasks and making the climb more hazardous. I did not realise how unfit I had become since leaving the Army, where once in those days a similar task would have seemed as nothing, today it was breathless and sweating that I reached the top of the stairwell, my only comfort was to note that Jim was in equally bad shape.

The most dangerous part of the job was climbing up the ladder to the maintenance ledge that circumnavigated the Lanternroom on the outside, once on the ledge you could secure the tether ropes to the handrail that was at chest height. I would scrape off any hard bits and wash the panes with soapy water; Jim would follow me rinsing the panes with fresh clean water before running a shammy cloth over them and finally with a damp cloth wipe the astragals, so his was the hardest job. That was not to last long, however; the alarm of the external bell of the telephone followed moments later by Iain's call had Jim as hastily as safety permitted, make his way down to answer the call. His promise to return to help me finish the task; were his parting words.

Something had cropped up which demanded his immediate attention, so he was unable to fulfil his promise. I did not mind in the least because it gave me chance to prove my worth. I was able to notice him scurrying about the Station and he did briefly stop to shout up to me that it was coffee time and I should finish the job after. It was indeed a timely call because the water needed changing anyway. I would not be going down with empty buckets though; I could not empty any dirty water onto the balcony for fear of staining the paintwork of the tower, likewise any attempt to throw the contents out as far as I could, would only result in getting it back. Even in what appeared to be calm at ground level the breeze up here seemed to be amplified and follow you in all directions. There was also the need to empty the tanks on the balcony that still retained the function of collecting water.

After coffee Jim was in the position to be able to help me finish the job but as I was about two-thirds of the way finished I suggested it was hardly worth the effort of getting back into his harness, Jim concurred but would help by emptying the balcony tanks for me. We had completed the job well before lunchtime but as we were both a little damp a change of clothes seemed appropriate. Before lunch Jim was to tell me what the telephone call was all about. Apparently the Fingal would be in the area at the time of our relief and Jim had to verify the state of the stores at the Station.

Both Principal Keepers completed a Monthly Returns Book which contained all the details of how much fuel and lubricants as well as notes on how many spare bulbs and any other equipment for the Station was on hand, he could have in theory just read the figures from the book but he was asked to verify in person their exact state to date.

The Station was fully equipped so was not in need of re-supply so The Fingal; just because she was in the area would make the relief when we went off.

Jim asked me if I would like to accompany him on a walk this afternoon, this certainly was a welcome change from previous experience and one that I was to enjoy on more than one occasion during the fortnight that I was there.

A herd of Red Deer came down from the hill to feed on the grass just outside the Station. This was the closest I had ever been to them in the wild and they did not seem to mind our presence so long as we did not make sudden moves or approach too closely. The herd was mixed with both stags and hinds that I thought unusual having heard of how aggressive stags can get towards each other during the rutting season. There did seem to be a dominant stag and it was apparent that no other stag was anywhere close to him, he had a bulbous lump that hung from his organ that I could only assume was prolapsed. It was also apparent that the hinds stayed their distance as well. What was noticeable at this range was that his and the other stags antlers were not fully developed and still had the greeny gray velvet covering on them ... On one of our expeditions Jim and I must have startled a group of four deer among the dunes. They took flight and three of them; led by a stag but not the one with that was prolapsed, fled into the Sound. The fourth turned the other way making its way up towards the hill. I wondered if the deer that had sought safety in the water would survive the current as they were swept further out into the Sound and carried at a rate of knots past the Lighthouse. Jim assured me that they looked healthy beasts so would manage well, in fact it was not a rare occurrence to see them crossing the Sound to get to Jura and vice versa. By the time we got back to the Station it would have been too late to make a courtesy call to the Coast Guard notifying them of the hazard to shipping and apart from the Ferry across to Jura some ten miles away the Sound looked empty of traffic.

Jim did give me some cause for concern on one of our travels when we came across a young adder sun bathing on the beach. I have a healthy respect for all nature especially animals that can harm us in retribution and snakes I have the highest regard for.

On one of the rare occasions when I caught my father in a talkative mood and when I'd asked the question of what my birth place was like; two stories he told will stay to the forefront of my memory: I was born in Kenya in a place called Gilgil where my father was stationed in the Army. Our married quarters were outside the Garrison and Pop once said that our garden was the whole of what is now Nakuru Game Reserve.
We had two native boys from the village that helped Mum and Dad around the house and our relationship with them was on the friendliest of terms, they referred to us as we to them by first names, the term Bwana was rarely used. Had it not been for the loyalty, trust and dare I say affection that they held for us, I doubt my father would not be alive today. My father had startled a black mamba that had taken refuge from the heat of the day under the steps of the wooden portico. Black mambas are notoriously aggressive as well as their bites being lethal and once disturbed can make quite a pace as they slither either in aggression towards or away from danger. On this occasion the snake chose the former and proceeded to give chase after my father as he ran for his life, the snake was followed closely by one of the boys who were thankfully armed with a machete that he carried to perform another task. Kenyans are noted for both their running abilities and agility and our boys were no exception. He won the race before the snake was able to catch my father and deftly beheaded the snake whilst at a sprint. This may not seem remarkable till it is realised that at the time of the incident the wind of revolution and resistance was in the air, the Mau Mau was making itself heard further south and more remarkable was that we were in the heart of Kikuyu country; homeland of Yomo Kenyata. It begs the wonder if the five pounds a month we gave the boys was being entirely spent on themselves and not funding an altogether sinister purpose and that their benevolent behaviour merely in protection of the goose the lays the golden eggs. My father insisted that there was genuine affection in the boys and they were as upset as we when we had to leave Kenya...

Jim was tormenting this poor creature to the point where I was sure that given the slightest chance it would have bitten him. I knew from reading that snakes are venomous at birth regardless of size or maturity and that even a bite from one as small as this one could prove fatal to anyone with intolerance to its venom. Although not overtly cruel Jim did prevent the snake from escaping by moving it back onto the open sand and I think it was this act that annoyed me as much as the fear for his safety. Eventually Jim succumbed to my protestations and allowed it to seek shelter in the grass.

On our other trips we would come across the antlers of deer; discarded after the end of the rutting season. Once we came upon an entire scull complete with antlers that Jim took back to the station knowing somebody who would give him a good price for them. Both Jim and I contemplated on the demise of this unfortunate animal and hoped it was old age rather than from combat or what seemed more unlikely a predator. It was not unusual for just the head of an animal to be found, all of the other bones could have been easily consumed and valued as a source of calcium so we did not scour the hillside in search of the remainder of its body especially as there was no value in it other than to shed further light on its demise.

Our furthest venture from the station took us just beyond the hill where the plateau of rock at the shoreline continued then disappeared about a mile further on. What lay rusting on the outer fringes was the hulk of vessel that had grounded many years earlier. There was nothing there worth investigating to warrant any disregard for the dangers that would be involved. Only rusting metal long devoid of paint remained, all traces of anything of worth had been stripped out long since. What drew our attention in almost utter disbelief was that out at sea just beyond the wreck and at some four to five miles was a Royal Navy Frigate speeding at some rate directly towards the wreck. It was as if the vessel was inextricably drawn towards the wreck in answer to some haunting plea or the wailing of some unheard and unseen Siren luring these mariners to their doom. We really became concerned for their safety should they continue on their present heading and speed and we prepared for the worst. At what seemed close to the last possible moment the ship turned sharply to port making a complete 180 before slowing down to a more sedate pace. This ship was to make two more similar manoeuvres before heading north and disappearing over the horizon. Had these manoeuvres taken place within view of the Lighthouse we would not have hesitated in raising a flag symbolising the letter U which stands for ... YOU ARE STANDING INTO DANGER ... from the flagpole.

The two weeks that I spent at Ruvaal seemed to pass so quickly, Jim and I would go for walks quite often and we both enjoyed the company. He would tell me of his exploits at some of the stations he had been to and the keepers he had met. The most interesting place he had been was InchKeith, that was his last Station. The island had a colourful history and he told me that bones had been discovered there that needed to be investigated by the authorities to discover their age and origins, investigation did not reveal all the facts but enough not to warrant further investigation because any culprit to any crime would be long dead. His relating the events to me was to have great influence on me at a later date when I recalled them and acted in similar fashion.

I hoped that the future would bring me once more into the company of this most affable man but for now I was happy that one more stage had passed in my progress toward appointment.

On the morning of the relief the Fingal passed the Lighthouse on her way to Port Askaig to pick up the relief Keepers. The Lighthouse flag had been raised and Jim had dipped the flag in salute as she passed. We had everything prepared for her return and it was just under the hour before we saw her turn to make her way back. Iain would remain at the station for another two weeks to complete his tour so remained standing by the radio in case the Fingal should call. Jim and I waited at the landing.

It seemed no time at all for the Fingal to come to a halt and lower her launch and even less time for us to be returning to it to climb the Jacob's ladder onto the deck of the ship. Our welcome and farewells for our relief were but a fleeting moment of that time. The ship had to turn again to make its way back to Port Askaig to drop us off before it would head further down the Sound to land stores and check the automated light at McArthurs' Head.

I was booked on the evening flight back to Glasgow, so Jim invited me to have lunch with his family at the Shore Station before I left for the airport. A car was waiting for us at Port Askaig to take us to Bowmore. At the bottom of Birch drive were four blocks of semi-detached houses all painted the same snow white with pale blue doors? Without doubt and because they were a different design and structure than the rest must be the Lighthouse Shore Station. As it turned out the first two blocks was the Shore Station for the Keepers at the Rhinns of Islay the second two for the Keepers at Ruvaal. I was impressed by the fact that they were all three bedrooms, not that we needed the extra room at the moment but a good idea for the future when the girls got older or should we decide to expand the family. Other than that there was not a lot to differentiate between them and other houses but enough to satisfy Margaret when I told her what she might expect from housing within the service.

Jim had arranged for the taxi to come at three o'clock and that time soon came, I said farewell to Jim and thanked his wife for the lunch. I wondered if fate would bring me back here to Islay, if not, whether I would meet up and serve with Jim again whenever and wherever that might be.

I had a delay before leaving the island; the flight had been over booked by six people, so six of us had to stay until a relief flight was arranged. I was only delayed by an hour and a half; it was one of the directors who piloted the aircraft, the embarrassment of the situation must have forced him into donning his captain's epaulettes once more.

I was to be home for another two weeks before I was called to go to Barra Head Lighthouse.

Extract from the Regulations...

SECTION J. WRECKS, MARINE EMERGENCY

J.1. AID TO SHIPWRECKED PERSONS.- Lightkeepers are required to give shelter and necessary aid to shipwrecked persons and any expense properly incurred in doing so may be notified to Headquarters for repayment. It follows that it is improper for Lightkeepers to claim or receive any remuneration from persons whom they have assisted.

J.2. CO-OPERATION WITH RESCUE SERVICES.- Lightkeepers are to co-operate to the fullest extent possible with the Royal Air force, Lifeboat, Coastguards and other rescue services, but such co-operation must never prevent them from carrying out their primary duties, namely maintaining the Light, Fog Signal, and Radio Navigational Aids installed at the Station. As a general rule communication with the nearest Coastguard Station should be established by telephone or radio on the following occasions:

(1) When any shipping casualty is observed from the Lighthouse

(2) When any ship or boat is seen to be in difficulties due to weather or any other cause.

(3) When any rocket or distress flare is seen.

Stations which have been supplied by the Department of Trade and Industry with white star rockets should in addition to communicating with the Coastguard, fire a rocket to acknowledge a distress signal and to indicate that assistance has been summoned from the Lifeboat and/or Coastguard.

J.3. DERELICTS, FLOATING OBJECTS.- Any derelict or floating objects observed from a Lighthouse which is likely to be danger to navigation must be reported by the quickest available means to Headquarters and to the nearest Coastguard Station. Such reports should contain as much information as possible about the nature of the derelict or floating object and should also include the force and direction of the wind.

J.5. Wrecks.- Principal Lightkeepers must immediately report to the Secretary and to the Coastguard any wrecks which may occur in the vicinity of the Lighthouse. The expression "wrecks" in this context covers not only stranded vessels but also cargo or other valuable material that may be washed up by the sea.

Reports are to be made by the quickest means-radio or telephone-and are to be followed up by a Wreck Return.

The Coastguard should be similarly informed if any person is seen to be interfering with the wreck.

J.6. SIGNALLING.- At certain Stations, which have been notified by Special Order, the Lightkeepers are to familiarise themselves with the rocks and shoals in the vicinity of the Lighthouse and are to ensure that if a vessel is seen approaching these dangers the appropriate signal is displayed namely flag "U" meaning "You are standing into danger". At night the letter "U" is to be flashed on the signalling torch.

Chapter VI
Barra Head

I would have to fly on this trip also but it would not be from Glasgow. I would have to get the train to Glasgow Central walk to Queen Street and get the train to Oban; I was booked into the Seaman's Mission for the night and then in the morning make my way up the quay to the NLB Depot where the helicopter would make the relief out to Barra Head.

If the train journey from Stranraer to Ayr seemed long; then the journey from Glasgow to Oban was an eternity. Something had caused a hold up at Springburn; it could not have been our engine because I could still hear it ticking over from my seat in the first carriage. Whatever the reason we were stuck there for over an hour, it was a good job that I had no other onward journey till the morning unlike some of my fellow passengers who were in fear of missing their ferries to Mull and other islands of the Inner Hebrides.

Once we started moving the journey was pleasant enough, with the sprawl of Glasgow behind us the Dunbartonshire countryside opened before us. Glimpses of the Clyde could be seen now and again as we followed its course to Dumbarton. Beyond Dumbarton the train took a spur that became single track and left the overhead electrified mainline behind. As the train climbed the gradient its speed began to slow and the labours of the engine increased as the throttle was opened up to maintain the momentum, it was at that moment that I wished this train had been one of the specials pulled by steam so that I could experience the thrill and the smell of the engines of my childhood.

Electrification did not come to the London Fenchurch Street to Southend Central line till the early sixties so as young boys my brother and I would watch the trains pass under a bridge that was about three quarters of a mile from where we lived. The bridge had a pedestrian footbridge attached to it and so we were protected from traffic. We would wait for a train to come and adjust our positions to match its direction of travel maximizing the effect from the blast of its funnel as it passed below us. The sensation as the slight tanginess and yet at the same time sweetness of the blast of smoke was almost intoxicating. The best effect was attained by trains on the down line, you could see them as they came round the bend and have time to adjust your position but the road part of the bridge masked its progress from view so you had to wait in anticipation. The sudden shock of the blast surprising yet adding to the thrill and the sensation of the smell that would linger for longer as the smoke trailed behind the engine for some distance. Trains on the up line could be seen from a long way off and their progress monitored right up until it passed below you, the sensation brief and fleeting for that moment only but worth it just the same. There was an alternative reason and a more hazardous reason for going to this particular bridge; underneath the wooden footbridge was a kind of trough that must have acted like a drain. In that trough were coins that pedestrians had accidentally dropped and had fallen through the gaps between the slats. Sometimes we would collect enough money for some sweets or a bag of chips, whatever, it was still worth the effort. Each expedition to the bridge would carry a penalty though, as we arrived home to the wrath of one or both of our parents as they remonstrated on our degrees of dishevelment.

The delay had meant that it was about half past five before the train arrived at Oban. It was only a short walk from the station to the Seaman's Mission once I had obtained directions from a bewildered local who thought I must have been blind not have noticed the large building with Mission to Seaman emblazoned on the front of it less than one hundred yards in front of me. I felt like saying that I was dyslexic but let it pass.

Next morning I walked the couple of hundred yards to the Depot, this was easy to recognise as it bore a striking resemblance to Stromness with the Buoys outside, plus I had recognised the familiar shape of the Fingal that was moored alongside. Just like Stromness: Oban was the Shore Station for several Lighthouses so there would be many reliefs today but as mine was the longest I would be the first. There were reliefs being made at Hyskier and Skerryvore as well as Barra Head but as usual before any of us could fly we had to watch the customary health and safety video.

The procedure was the same as going out to Sule Skerry and I acted like an old hand even though I was the only one travelling to this Station. The flight to Barra Head was about forty-five minutes and took us over the Isle of Mull; once we passed Mull there was nothing but the Minch; the stretch of water between the Inner and Outer Hebrides.

At three thousand feet the view was intermittent as we entered cloud but I was able to pick out a group of islands ahead of us in the distance. We appeared to heading for the southernmost and smallest of these islands. Barra Head was not as I first thought on Barra itself but was an island at the southern tip of the chain some fifteen miles or so from Castlebay.

I could see as we approached the Lighthouse that it was in an elevated position at the top of this island, the altimeter dropping below the one thousand mark as we made our landing. The cluster of buildings that formed the familiar pattern of Lighthouse Station was this time only barely dwarfed by the short stubby tower. What need of a tall tower at this height above sea level?

The helipad was just inside the perimeter of the boundary walls and about thirty yards from the buildings, I had expected to see a trolley waiting for us to put my gear and the provisions on but a tractor and trailer were parked just below and clear of the helipad. I was wondering whether I had misjudged the amount of stuff we had loaded into the back of the helicopter or there was something I had missed seeing on our approach but when I noted the eccentric looking chap approaching us; it all seemed to fit into place. After unloading the helicopter and bidding farewell to the chap I was relieving, it was moments and the helicopter was airborne again. Once the noise of the helicopter disappeared the eccentric character introduced himself as Alan Chamberlain, the best was way to describe Alan would be to liken him in appearance to John McCrirrick of racing commentary fame. I'm not sure if I am remembering him wearing a dear-stalker or confusing him with John, however he did have the same bushy side-burns and glasses and the same kind of eccentricity in his bearing but outwardly he was more jovial and affable, as he always seemed to have a smile on his face.

Barra Head Lighthouse was almost fully automated the only thing was the linking of the monitor to the system being established in Edinburgh. I began to feel that I would not get become appointed as Assistant Lightkeeper before all of the Service was automated, luckily for me as Alan explained showing me the details of the planned progress for automation written in the Lighthouse Journal the bi-annual magazine printed by the Board. As there were still over sixty manned Lighthouses left the odds were heavily in my favour and I felt comforted by that reassurance. So once again there would be little to do and nothing to learn in the way of Lightkeeping at this Station but Alan would always find us something to do in maintaining the Station.

Whereas Sule Skerry was the most remote Lighthouse in Britain; Barra Head was the highest. It was no wonder that there seemed little point in having a fog signal here, most of the time the fog signal would be on as for hours on end the Station would disappear under a veil of cloud and its sound would probably be inaudible at sea level as it was carried away. I could only imagine the efficiency of the light at night under the veil of cloud but as the Lighthouse had cast its beam for nearly a hundred years it must have proved its worth to mariners.

Alan was the 1st Assistant and the responsibilities of Principal Keeper fell on him in his absence and it was a mixture of Alan's eccentricity coupled by his enthusiasm to find things for us to do that led to my stay at Barra Head being interrupted by the necessity of having to visit the Doctor in Castlebay. I cannot lay the blame fully on Alan's shoulders because I had been suffering the effects of a heavy cold since contracting it a couple of days after my arrival, however the chain of events that followed will always be remembered as a comedy of errors.

The Island rose from sea level at the landing to nearly seven hundred feet at the Station about a mile or so away, at the landing was a shed that contained a half dozen or more cement bags that were passed use. Instead of slipping the bags into the sea at the landing Alan had the idea that we should take them up to the Station and toss them over the wall, whether this was for the benefit of entertainment as they plunged dramatically onto the rocks below or in an attempt for the bags to break open and fragment as they crashed their way down onto the rocks therefore dispersing the contents over a wider area, I had neither the inclination or well being to ask. My concern was that my strength seemed to wane as every bag was tossed over the wall to the count of... One ... Two ... Three ... Heave. When the penultimate bag came into our hands my strength seemed all but spent; on the count of heave the bag barely manage to clear the top of the wall and as a consequence the nail of the middle finger of my left hand caught the coping of the top of the wall. Any discomfort I felt from my cold was superseded by the immense pain that I felt now. Blood dripped copiously from the injury or as it would seem to any injured party, relief was sporadic as the pain came in waves only by holding my arm up and gripping my left wrist hard diminished both the pain and the flow of blood, this reaction was more instinctive than rehearsed or instructed, however Alan guided me toward the kitchen and living room where I could get the wound cleaned assessed and dressed.

The pain eased as my finger was run under the tap to remove the layer of cement dust that coated my hands. It became apparent that my nail was still intact but barely attached to my finger. Alan was concerned that it might become infected and felt the need to seek advice from the nearest Doctor, emergency contact numbers were listed in the radio room as was the medicine cabinet so it was there that we made our way once the wound was clean. The advice given by the Doctor was to verify when I last had a tetanus injection and being assured that it was within the last five years was satisfied that I should be covered from that type of infection however the nail should remain as is and not be removed, the wound should be cleaned and dressed with a gauze and bandage and pain killers in the form of codeine should be administered every four hours. Should I continue suffer the same extent of pain in the next couple of days then he would be advised to contact the Doctor again.

Alan appeared to have the confidence and know what he needed to do, by this time I was not in a fit state of mind to notice let alone question his actions, so I failed to notice that Alan had left the protective wax covering on the gauze when he bandaged my finger. I was also under strict instructions not to interfere with the dressing for at least forty eight hours by that time the healing process would be well under way.

The pain had subsided as the codeine began to take effect, but by morning my finger began to throb constantly the codeine having no effect at all. I suffered the pain for one more day before finally agreeing for Alan to seek the advice of the Doctor once more. This time I would need to go ashore for him to see the wound for himself before judgment or diagnosis could be made, so Alan had to call on the boatman from Barra to come and pick me up, he also had to notify the Board of the situation and for them to be satisfied that the right decisions were taken. However once I was ashore I would have to call them myself and let them know the outcome.

It took an hour or so for the small fishing boat owned by Hugh McNeil to arrive, with the aid of binoculars we could trace his progress from eight miles away so had plenty of time to get down to the landing. In case of the worst scenario I had to pack my bag though at the time I was sure that I would return before nightfall.

It was nearly midday by the time the boat landed at the jetty in Castlebay. The way that Hugh tied up his boat had the kind of permanence about it that begged me to ask the question about my returning to Barra Head. "Oh wee'l no be going back there today" Hugh replied ... Shit... I thought ... What the hell have I got myself into ... The next sequence of events seemed to have the air of conspiracy about it, when I arrived at the surgery the Doctor was on his rounds and would not be back till evening surgery at five thirty. So I had nearly five hours to kill before I could see the Doctor but whatever the diagnosis I needed to let the Board know that I would not be getting back to the Lighthouse today unless they could persuade the boatman otherwise. I waited at the phone box for them to return the call once they had spoken to the boatman. After about fifteen minutes it rang and I was relieved to hear that it was the Secretary himself on the other end. I was informed that I would not be going back tonight and therefore accommodation had been booked for me in the Castlebay Hotel ... which co-incidentally happened to be owned by a Mr Neil McNeil ... I was to phone again and leave a message if need be once I had seen the Doctor, should anything else occur over the weekend, I was to phone the duty officer who's number would be left at the hotel. So that would be my first call, trying to find the hotel which would not be too difficult as the most obvious choice of building stood higher than its neighbours and had every bit the look of hotel about it.

There were lots of things that set Castlebay apart from other places I had seen; for one thing there was a sense of detachment as each dwelling seemed to occupy its own space and there was very little evidence to show that space was at a premium, only in a couple of places were there semi-detached houses or the slightest hint of terracing in the form of shops and business premises that had somehow clumped together around the harbour. Out in the centre of the bay was the Castle that gave name to this place. It's very shape indicative that this was not merely a fortress to defend its occupants from an attack from the sea but latterly; as the dwelling inside its bounds had taken on a more modern form, most of its windows overlooked the town for the clan chief to observe his subjects.

I dropped my bag off into the room I was allocated, taking note that the Board were going to pay for all my meals as well. However they did not do lunches during the week, only at weekends but they would gladly make up some sandwiches for me. By the time I had finished unpacking and made my way back to the desk my sandwiches were ready. There was only one thing left to do and that was to somehow let Margaret know what had happened, I had left it till now thinking there was no need to unduly worry her. It was difficult to make contact with Margaret, as we had no telephone so we spoke on a weekly basis when she went to her Aunt Jessie's with the girls. It just so happened that Margaret had planned to take the girls and spend the weekend at Jessie's so there was every chance that she might already be there or would arrive there shortly. As it was she had not arrived when I called so I left a message with Jessie, without telling her my plight, and that I would call later in the evening.
... Jessie was more like a mother to Margaret, even before the death of Mary some four years earlier. After her death both she and Tam, took on the mantle of parents for Jennifer; Margaret's much younger sister and the bond between all of them grew till it became hard to realise their relationship was anything less. Our own children and grandchildren still regard her as Nanna or more affectionately as she preferred just Jessie. She had to find the strength to continue after Tam's death which sadly deprived us all of this much loved affable gentle and affectionate man; she did so by devoting her time by mothering all of us as well as caring and being neighbourly to others around her. She did therefore tend to worry more about others than herself, so I needed Margaret to tell her that there was nothing to worry about and that I would be fine, that of course depending on my being able to convince Margaret that I was OK ...
In the entrance hall of the small hotel was a map of the island and apart from a couple of spurs that didn't amount to much; there appeared to be only one road that circumnavigated the whole island and that I should manage the eight to ten miles indicated by the legend quite comfortably and still have time to spare for my appointment.
The island was dominated by the hills that joined together to form one at the centre of the island, prevented this observer from seeing any more of the coast than what was immediately in front of me. Granted I could see some of the other islands that formed the chain but once they passed from view there was nothing but the North Atlantic, I dare say that if I had climbed to the top of the highest point it might just have been possible to see the islands of the Inner Hebrides on the horizon but at just above sea level there seemed nothing but the emptiness of the ocean. Apart from the occasional bungalow or hamlet where a few houses had grouped together there was little else in the way of habitation. I could see there was nothing on the scale that could call itself a farm as on the mainland but where some of the dwellings had what seemed a surplus of outbuildings and there was livestock close by and on the hillside, I took these to be the crofts that I had once read about. Somewhere at about half distant, for I had now lost track of how far I had come; was a sign saying Airport but it failed to mention how far it was so I did not take up what might have been an invitation for further investigation.

I would not have missed anything at this particular time of the day but had the tide been right, I would have seen the wondrous spectacle of an aircraft landing on the beach that acted as the runway of Barra's Airport.

My legs and feet began to weary the further on I walked, at about three quarters of the way round a red post bus pulled up alongside and asked me if I wanted to get on and save my legs ... no doubt the island grapevine had sprung into action, figuring that I might not last the course and McNeil the Post hoped to he might get a bit of the action. .I think it only fair on the islanders that my judgments may have been tainted by watching too often the films Whisky Galore and Rockets Galore which I understood had been filmed on location on Barra and that the course of these current events were merely circumstantial. Whatever, like the trek to Loch Doon, I was determined to see this one to the end so I declined his offer.

I kind of figured that the chain of islands that formed the group ran approximately north to south and there were no islands close by to the east or west so by that reckoning the partial outlines of the island that I could now see before me was the same island I could see from Castlebay and if true, my walk would soon be at an end.

The relief of entering the hotel and smelling the rich aroma of freshly made coffee being carried on a tray into the guests sitting room intimating that I was not the only person staying at the hotel and that I too might sit and relax my weary bones and refresh myself with a cup without the need of climbing that stairs and making do with the instant variety that adorned the table in my room.

The Doctors surgery was empty at half past five so I had not long to wait before being seen. My finger had throbbed the whole time but I was occupied by other things and the codeine did lessen it to a degree. "Let's have a look at this finger then" he said, the bandaging and the phone call from Alan negating the usual "What can I do for you?" He tutt-tutted more than once before finally he unstrapped the bandage to reveal the gauze below." Who on earth dressed this he said"? A suspicious glance into my eyes and the innocent expression my face was enough to convince him it was not my doing was soon followed closely by another series of tutts that eventually concluded the interrogation.

When he gently lifted the gauze and my finger was released from its strangulating grip the relief was both immediate and enormous ... The protective strip had contracted over the past couple of days, pushing the nail against the exposed nerve endings thus causing the persistent throbbing ... He never said as much but I'm sure would have agreed with my diagnosis, his only comments were that "It seems to be healing well" and "There doesn't seem to be any infection, but I'll give you another Tetanus jab just to be on the safe side," however they did little to persuade me that he could not but wonder what all the fuss was about. He redressed the finger and before allowing me to replace my arm in the sling; gave me a jab in the same arm, giving my nervous system yet another working out till the pain of the jab subsided a few minutes later. I could do no more but thank him and express what a relief it was now that my finger had been dressed properly and apologized for any inconvenience that I may have caused. "Save your apologies for the Board" he replied, this reply left me wondering if he had more than just a passing interest in the Northern Lighthouse Board and that I would not be the only one reporting on the outcome of my visit.

The next morning apart from being a Saturday it was also stormy. The wind had got up during the night to peak in the early hours to gale force. There would be no ferry from Oban and likewise there would be no boat to take me back to the Lighthouse. Even if God was to abate the storm by the end of the day the islanders were strong believers of upholding the Sabbath so would not dream of taking me back on a Sunday so it looked as if I could find myself stuck here till Monday at the earliest. I now had the task of notifying the duty officer of the situation. Things were now well and truly beyond my control and the most frustrating part of the whole thing, I was carried along by events in which I could only be a passenger.

The wind and swell abated by late Sunday afternoon and by Monday morning the ferry service had resumed as normal. I had anticipated being able to get back to the Lighthouse by the afternoon but felt thwarted once again to be told that the swell was still dangerous for such a small boat. When I phoned the Secretary to once again update him, I suggested that should I not be able to get back today or by tomorrow morning I was might just as well get the ferry back to Oban as I was due to be relieved along with Alan on Wednesday in any case.

He must have able to apply some pressure because by afternoon Hugh had relented and deemed it possible for me to get back before four o'clock True to his word we landed at the Rock at ten minutes to four; ending the saga of the sore finger.

I refrained from letting Alan know of the Doctors disdain at Alan's nursing abilities, however, I did let him know of his mistake on not removing the protective coating on the gauze before applying it, only to prevent the same series of events from happening to some other poor soul in the future.

The return flight back to Oban seemed to take longer, mainly because we were flying through unbroken cloud throughout and had nothing to observe and because journeys home always have that effect.

Margaret was happy to see me home again even if I was a little more damaged that when I left.

The Board must have been happy that everything that happened was due to the accident of losing my nail and that what followed was just circumstantial. Their main concern now was that I should notify them when I had a Doctors opinion that I was fit enough to return to Duty and that I should not seek his opinion for at least a fortnight.

It would be late August before details of my next assignment were delivered, this time I would be going to Holburn Head to relieve the Principal Keeper who was going on Holiday for two weeks.

Extract from the Regulations...

Section L. Medical

L.1. GENERAL.-At all Light Stations a stock of medicines and medical stores is to be maintained in accordance with instructions sent from the Secretary. At Rock and other inaccessible Stations the medicine chest will contain a comprehensive stock, while at Stations where a doctor is easily obtainable; a smaller quantity of medical stores will be kept.

All Lightkeepers must register with a doctor in the vicinity of their Light Station or Shore Station. Any Lightkeeper requiring advice in this respect should refer to his Principal or communicate with the Secretary.

L.2. DENTISTRY.-The Commissioners expect the Lightkeepers in their Service to take proper care of their teeth and to make a point of registering with and visiting a local dentist at least once a year; this is particularly important when serving at a Rock or other inaccessible Station

A Lightkeeper suffering from toothache when out at a Rock should request his Principal to get in touch with his local doctor or dentist through the Shore or Control Station R/T.

In exceptionally severe cases the Lighthouse Tender, Helicopter or Attending Boatman may be sent out with a relief but the Commissioners would not regard a relief of such a nature with favour if the trouble could have been avoided by regular visits to the dentist.

L.3. First AID.-Pamphlets and posters on first aid will be distributed to Lighthouses from time to time and it is the duty of Lightkeepers to ensure that they are familiar with them particularly with the methods of artificial respiration.

L.4. MEDICAL ASSISTANCE.-In the event of sickness or injury at a Light Station, it is the duty of the Principal Lightkeeper to obtain medical assistance or advice without delay. As a general rule this will be a matter of telephoning to the local doctor. The telephone numbers of doctors with whom Lightkeepers are registered should always be readily available.

At Rock Stations a radio call should be put through to the Shore or Control Station at the first opportunity and the symptoms or nature of the sickness or injury described to the Lightkeeper ashore, whose duty it is to communicate this information to the local doctor and to inform the Secretary. The Rock Station is to continue calling the Shore or Control Station frequently (that is to say, at approximately 10-minute intervals) until the doctor's instructions have been passed. Depending on the doctor's report the Lighthouse Tender, Helicopter or Attending Boat may be sent to carry out a relief. If medical assistance is urgently required the Rock Station may also call the local Coast Radio Station to request medical advice.

SECTION N. ATTENDING BOATMEN

N.1. GENERAL.-Attending Boatmen are at certain Rock and Island Stations to undertake Relief Trips and other missions as may be required. They and their boats are to be at the service of any officials from Headquarters, or any other duly authorised person.

Where the boat is the property of the Commissioners it is to be employed on lighthouse duties only. But nothing in this Regulation prevents the boat being used for lifesaving or other local emergency work, or as the Marine Superintendent may direct.

Where the boat is not the property of the Commissioners and is regularly used to transport Commissioner's personnel, safety standards as required by the Marine Superintendent are to be observed; the boat may be inspected periodically by the Marine Superintendent or his representative.

N.2. ACCOUNTS.-Where an Attending Boatman is paid by the trip, unless instructions to the contrary are issued from Headquarters, he is to render to the Principal Lightkeeper in the first week of January, April, July and October each year an account for the trips performed in the previous quarter; extra trips and the reason for them are to be clearly shown on the account.

Where the boats crew are paid by the Commissioners the account is to be accompanied by a Pay Bill which is to be certified by the Principal Lightkeeper.

N.3. CONTROL OF ATTENDING BOATS.-The control of the Attending Boat so far as is not provided for in this section of the Regulations is to rest with the Principal Lightkeeper, who is authorised to give the Boatman any reasonable instructions for attending the lighthouse having regard to the weather and landing conditions.

Boat trips to and from the lighthouse are nevertheless at the discretion of the Boatman, who is responsible for the conduct and safety of passengers and crew and may give any proper instructions to them to ensure the stability and seaworthiness of the boat.

Only Boatmen duly authorised by Headquarters (Marine Superintendent) are to be in charge of the Commissioners' Attending Boats.

Chapter VII
Holburn Head

In terms of time spent travelling my trip up to Holburn Head would eclipse by some four hours any journey previously taken at the behest of the Board. During its course I would need to change trains three times and have the walk from Central to Queen Street Stations. I also had to make sure that I was in the right section of the train from Inverness to Thurso/Wick; the front half of the train would de-couple at Georgemas Junction and go on to Wick; the rear half would be coupled to its engine and go on to Thurso, thankfully I had observed that the carriage I was in had a sticker on the window indicating that I was in the right section but nevertheless I had this confirmed by the ticket inspector as he hurriedly passed by during the de-coupling process at Georgemas Junction.

I arrived at Thurso at about half past eight and had to wait in the queue for a taxi. I was about fifth in line and it seemed to be the same three taxis that were operating so it estimated that it would be nearly ten minutes before it was my turn, or so I thought, a different taxi appeared at the stance to take me to the Lighthouse that was three miles from Thurso across the bay. The talkative lady driver had me relate to her my reason for going to Holburn Head and noting my accent asked me in ignorance if I had travelled all the way up from London. Her ignorance lay on two counts, one she could hardly have known that I had lived in Scotland for two years and two The Northern Lighthouse Board is completely independent from Trinity House who govern the lighthouses of England so would be unlikely if ever, to call upon the services of one of its members or employ and pay for somebody to travel such a distance.

She was not the only one to be touched by ignorance because I mistook her Caithness accent to come from either Devon or Cornwall; eventually we educated each other before our journeys end.

Alec McKay was the Principal and only Keeper at this One-man Station. The Station was at the end of the road that went up the hill beyond the Harbour and Port of Scrabster; this was home for the Ferries across to Orkney. It was not quite on the Headland at the end of the arm of the bay, unlike the Lighthouse of Dunnet Head that stood proudly on the cliff top at the end of the opposite arm with Thurso itself between the two.

The lighthouse had an outward appearance more like a church but with a tower instead of a steeple and the house that it was attached to with its two storey windows was for living in rather than single stained glass windows of a house of prayer. This time the facings and corner stones were painted a slate gray not sandstone but still unmistakably Lighthouse just the same. This however was not where Alec stayed, his modern detached house was just inside the entrance gateway, and it was from the side door of this dwelling that Alec appeared to bid me welcome. This time I would settle up with the driver myself as it would be me who would be sending in the Returns for this Station at the end of the month, the Board had kindly sent me a receipts book; it was pocket sized and had a black hardback cover, each cream coloured page apart from the inside and last page; bore the Board's logo. Without so much as saying so the Board had presented me with a second clue that my days as a supernumerary were numbered, the first of course was the added responsibility of as acting Principal Keeper.

The couple that occupied the house attached to the tower was a Lightkeeper called Colin McEwan and his wife. He was a Keeper who was at that moment out at Stroma, he may have been the Keeper that the helicopter went to Strathy Point to pick up; I never got round to ask.

Alec showed me to the Bothy; a small single storey dwelling that sat between the two. This was a bit larger than the one at Killantringan for it housed the semblance of a small if somewhat compact kitchen. Apart from that the layout was much the same.

I would be given a day's grace to settle in before taking up my duties; this would be ample time for me to settle in and go into Thurso to buy provisions to last me till my next day off. Once I had decided what day I preferred to have off I was to call upon the Occasional Keeper, Will Sutherland who would relieve me. Alec suggested that a Thursday or Saturday would be good days to have off, as it was what he had been used to. I felt there was no need for me to change anything that might spoil their working relationship; after all I was merely custodian of the position till Alec's return.

As it happened Alec would still be at the Station for another five days before going away anywhere. If there were any problems he would be on hand to give his advice if I needed it. There was little for Alec to show me in regards to my duties; the light was turned on by a switch and it was the lamp that flashed not the rotation of the lens so there was no winding gear, the fog signal was an electric emitter that also had a switch and apart from general cleaning and maintenance of the Station and grounds that was all there was. I was not expected to do anything more than at other Stations I had been to and intended to observe the routines that I had learned so far, the only exception was that with regard to the fog signal; I would be advised not to wander too far from the Station if there was the likelihood of fog and at night the Coastguard would wake me by telephone if there was any fog.

I did ask Alec if it would be all right for me to bring Margaret up for a week, his answer was that it was not his decision to make, as he was not the Principal Keeper but if I was to ask his advice, he saw no harm in it.

This would be a golden opportunity for Margaret to get a feel of what we might expect once I was appointed. Reading between the lines prior to my assignment here I had intimated to Margaret that should the chance present itself I would if she wanted bring her up to Holburn Head for a week. So all that was left for me to do was to phone and ask her.

I past my first few days busying myself with the routines; in the afternoons I felt at a loss for things to do, there were quite a few tourists and visitors who came up to the Station or took the coastal footpath behind the Station to the headland beyond. A sign on the gate notified visitors that they may at the Principals discretion be admitted to the Station between the hours of two pm and an hour before sunset. I thought this might be a good way of passing the time. I felt a little nervous that they might ask questions that I might not know the answer to. However most of them found the experience an interesting interlude before moving on to some other form of attraction. For some it was a pastime waiting for the ferry before making their onward journey to Orkney.

On Sundays I would fly the Lighthouse Flag, which seemed to attract even more visitors than normal; drawing them to the Station like a sign that says "Open for Business". I would sadly have the need to fly the flag on other occasions during my stay at Holburn Head. Just before Alec went away I had heard on the news of the tragic death of Lord Louis Mountbatten. I felt compelled because of the affinity between the Service and the Royal Navy; Lord Louis at the time also having the title First Sea Lord, to fly the flag at half mast out of respect. So I did not require the instruction from the Board to do so however belated that message arrived. A letter arrived from the board also notifying me that the Royal Yacht would be arriving at Scrabster with the Royal Party. This was an annual occurrence but may have been brought forward by the death of Lord Louis. The Queen Mother had a residence at the Castle of Mey just up the coast from Dunnet Head and the family I believe were in residence. Whatever the circumstances; I was to be aware of the possibilities and make myself available at the station to offer the Services courtesies on their passing the Station.

The grapevine in the form of the Coastguard would notify me when the Royal Yacht passed a certain point, but other grapevine sources let me know a day in advance of when she was due to arrive, so I did not have to hang around the Station all day every day. Luckily, Margaret had arrived a couple of days earlier so could watch the spectacle. The day did not start off very well as it was both wet and windy. The wind was strong enough for me to warrant putting up what I mistakenly thought was the storm flag; the problem with this flag was, that although smaller and less likely to be affected by the wind, it did not have the sturdy ties to bend it onto the halyard.

When the time came I would lower the flag to just above half way and the raise it again on seeing the response from the ship. The response I was looking for would be for the ensign on the sternpost to be lowered and raised in acknowledgement of my salute in the case of the Royal Yacht or; as was the case, the Yacht being preceded by a Royal Navy Frigate who flew ensigns at both the sternpost ensign and the mast either of which could be dipped.

I judged the timing just as the Frigate's bow came level with me, and began to raise the flag again once I noted her response. Just as the flag reached the top, the bend I had used became untied and the flag fell limply to the mast and then to the ground. I could just imagine the Captain's expression of seeing such a shoddy display, but that was to be the least of my problems. The Royal Yacht was less than a mile away and I had the major task of retrieving the other end of the halyard, which was stuck at the top of the mast. If I did not get the flag up in time to salute the Royal Yacht then I would surely incur the Commissioner's greatest displeasure.

Luckily for me there was a boat hook in the shed, to this day I can only but wonder why a boat hook would be at this station especially when there was no landing. Alec could only tell me that it must have been there for some time because it was there since before he came. Whatever the circumstances it came to my rescue that day even if I had to extend its length with wire in order to reach, however it was a very close thing to retrieve the halyard re-tie and raise the flag before the Royal Yacht's bow drew level. In what must have appeared all one motion the flag was raised and lowered to its dipped position and no one was more relieved than I to see the sailor at the stern post acknowledge my salute.

Thankfully for all, the wind seemed to die down, the sun came out and it was to turn out to be quite a pleasant day. The Royal party were to stay for a few hours before boarding the Yacht once more. Giving me time to exchange flags for our larger and more traditional one. I gained a valuable lesson on that occasion, it would be far better to risk having the flag torn to shreds than to put up a flag whose ties and my skill at knot tying were not equal to the task.

The Royal Yacht had three escorts: two frigates and a destroyer. The Frigate that led the Yacht took station just at a point beyond the harbour; the other Frigate took station at the mouth of the bay and the destroyer at the entrance to Dunnet Bay.

The Royal Yacht flew three Standards at her masts so there were two other members of the Royal Family beside Her Majesty on board.

Margaret could have had a better look if she chose to go down to Scrabster but she chose to stay at the Station with me. Colin was home from the rock and he and his wife joined us to watch as the Royal Yacht prepared to leave Scrabster.

I would not have to dip to the Royal Navy Frigate this time as she was beyond the two mile acknowledgement limit and she was travelling at a good rate of knots in order to take up her station at the Head as the Royal Yacht came out of the bay into the Firth. All passed as planned, I dipped the flag as her bow came level and raised again once I had seen her response. Colin did question me as he said he did not see her respond. Both Margaret and I clearly saw the stern ensign dip in acknowledgement so I wondered if Colin was looking at the right ensign. As I had not suddenly taken on the status of Royalty that remotely warranted an acknowledgement from one Monarch to another, so he could hardly expect to see the Royal Standard being lowered in response. I never pressed him further and was happy that I had fulfilled my duty.

Margaret and I decided to go to Wick on my day off, so I duly asked Will if he could relieve me on that day. Will though short in stature was big in heart and would pleadingly ask for me to give him more days than the Board would sanction, he even offered to do the relief unpaid, which although kind of him I sadly could not oblige. He lived in a terraced cottage in what affectionately was called the fisherman's row; it was a small one bed roomed dwelling similar to some of the miner's cottages that our locals called but and bens. Now if anyone was to epitomize the character of a Lighthouse Keeper or that of a lifetime spent by or at sea then Will suited the bill perfectly. Into his eighties he liked to keep himself busy but had the air of loneliness about him. His affability and hospitality at times seemed overwhelming as Margaret and I found on more than one occasion. Once while we were taking a stroll into Scrabster we were passing his cottage when he appeared at the door and would not take no for an answer as he shepherded us into his living room come kitchen. The smell of fried fish permeated the air, and without further preamble he announced, "Do you like fish?" this was not really a question, it was more like an order, as he began to plate up two dabs that he was frying in a pan. But we could never the less do no more than return our affection for this loveable character.

It was soon time for Margaret to return home, we had had a lovely break together in what at most times had the holiday feel to it. Margaret's maternal instincts were kicking in and she missed the girls as did I but it would be another week before I would be home.

The view from the Bothy overlooked the bay and I would spend hot afternoons soaking up the sun and taking in the scenery. Some days in the early mornings or late evenings the salmon netters' would come to check their nets that were set sixty or so yards from the lighthouse. Sometimes days would pass before the nets would yield a catch of either a sea trout or a salmon. Will had put me off the idea of fish for a while or I might have been tempted to strike up banter with the netter to get a discount on a fish. Talking of fish, I was privileged to be in Scrabster at the time when a large halibut had been landed that took the record at the time for being the heaviest fish caught in British waters.

One Sunday afternoon, there was a spectacle that seemed to follow the script of a Soap Opera. Every Sunday another ferry called the Smyrill would dock on her weekly voyage round the North Sea to the Baltic. It would take a week for her to go from Scrabster to Lerwick, the Pharoe's, and Norway, Sweden and Denmark and back again. However on this occasion she was late in arriving at Scrabster. You could sometimes see and imagine the frantic efforts to get the ferry unloaded and loaded before the Orkney ferry whose berth she occupied was due in. I could see the approach of the ferry into the bay and could only sit back and watch as the drama unfolded. It was like watching two unruly children bickering when one had occupied the others seat. The entrance to the terminal was only large enough to handle one ferry at a time and only had the facilities for one. It was obvious that the Smyrill would not be able to complete her loading and as the terminal was in the ownership of P & O, they would have to give way and complete their loading once the Orkney ferry had left. The pandemonium could be observed from here as the bustle of cars and Lorries jostled for their new positions in line, the sound of frustration as drivers beeped their horns disapprovingly, eventually calm was restored and both ferries departed huffily on their way.

In a way I was sad to leave Holburn Head because it was a place that I began to feel like I belonged and it had the sense of home about it, however there was something that had an even greater pull besides the desire to get home to be back with the family once more; I had received a letter from the Board notifying me that it was with their pleasure that I was to be Appointed Assistant Lightkeeper on three months' probation at InchKeith Lighthouse. At last I had been fully accepted into the Service and was no longer going to be a SPARE MAN.

Extract from the Regulations...

SECTION K.- PAY, ALLOWANCES. ETC

K.1. ACCOUNTS.- In addition to statements of travelling expenses (Regulation K.11), Principal Lightkeepers should submit accounts in respect of:

(1) *Station payments (Station Accounts)*. These are to be forwarded to the Accountant on the first day of each quarter to cover the previous quarter and should include, where applicable:-

- (i) Rock Allowances
- (ii) Payments in respect of laundry
- (iii) Payments to Attending Boatmen
- (iv) Authorised car hires (church, provision Trips etc). When payment is to be made locally

(v) Work authorised from Headquarters (supporting authorisation to be attached).
(vi) Carriage and freight accounts.
(vii) Postages and similar incidental expenditures.
(viii) Tradesman's accounts in accordance with Regulations H.8. And K.10.

(2) *Occasional Lightkeepers.* These accounts are to be submitted monthly, and must state the name of the Occasional Lightkeeper, the number of days duty at The Lighthouse and the reason for attendance.

K.11. TRAVELLING EXPENSES, SUBSISTENCE ALLOWANCE AND REMOVAL ALLOWANCE.- Lightkeepers transferred from one Station to another are entitled to travelling expenses and subsistence allowances for themselves and their families (children up to the age of 18) and a removal allowance for alterations to curtains, carpets, etc, may be claimed after arrival at new Station at the rates notified from time to time by Headquarters. Claims for payment of these expenses and allowances are to be submitted to the Accountant on a Travelling Bill at the first opportunity. As a general rule no allowances are payable in respect of periods spent on a Lighthouse Tender.

Lightkeepers are also entitled to submit claims for payment of their own travelling expenses and for subsistence allowances in respect of any journey undertaken at the behest of the Commissioners.

Lightkeepers and their families leaving their Station on retirement are allowed to reasonable travelling and removal expenses but no subsistence allowances are payable.

A Lightkeeper dismissed from the Commissioners' Service or resigning of his own accord will not be entitled to any travelling expenses or allowances.

K.12. TRAVELLING - FURNTURE AND EFFECTS, MOTOR VEHICLES.- The Commissioners will meet reasonable costs of freight and carriage charges on Lightkeepers' furniture and effects when moved on transfer from one station to another. Motor vehicles should normally be driven to the new Station and a mileage allowance may be paid for such journeys provided the vehicle is owned by the Lightkeeper or his wife. When the vehicle is to be transported by sea the Commissioners may meet reasonable freight charges incurred if it is not possible for the transport to be made by Lighthouse Tender. In all cases of a car having to be transported by sea the Lightkeeper concerned should apply to Headquarters for permission to ship it by Tender.

K.13. TRAVELLING - INSURANCE.-The Commissioners will reimburse the cost of insurance coverage on the furniture and effects moved contrasted up to the value shown in the letter of transfer. On request to Headquarters the Commissioners will also provide comprehensive insurance coverage up to £800 on cars belonging to Lightkeepers or their wives when on transfer involving sea journey whether by Lighthouse Tender or on ships owned by commercial shipping companies. This coverage will be limited to the period when vehicles are in the process of loading and unloading and in transit by sea. Insurance coverage from Lighthouse to point of embarkation and from point of disembarkation to Lighthouse will continue to be the car owner's responsibility.

Chapter VIII
While you were away
(Wait till your father gets home)

I am happy that Peter has thought it important to include my side of the story, for me the whole experience and way of life meant the world to me and offered to make those years the best and happiest of my life.

The title of this section of the book was chosen unanimously by our children who when asked for their suggestions for a title for this chapter all came up with this. The kids are not really being fair on me because ... Wait till your father gets home ... could only be called my catch phrase while Peter was away at the Ailsa Craig and I had the three of them to deal with. If you have ever seen the cartoon of the same name then Peter would definitely be a ringer for the father. He was not the ogre I sometimes painted him, but his words would be enough for the children to behave instantly. The kids said I used this threat constantly along with other methods to control them but I think they are just exaggerating.

We have many memories of our time in the Lighthouses, so it takes more than just one chapter. So as we travel from lighthouse to lighthouse I hope that you might get to see what it was like for us; the wives and children to be part of the Lighthouse service.

When Peter first showed me the advert in the paper, I could see that he was excited and I got taken in as well. Ever since I was a little girl I had looked at a plaster of Paris souvenir of a lighthouse on our mantelpiece and wondered what it would be like to live in one, now I might get the chance.

I was born and brought up in Burnton which had a separate identity from Dalmellington even though it only consisted of a few rows of miner's cottages, a working men's club and what used to be a store but was later turned into another pub.

When we moved back to Scotland we could only get a house in the scheme in Bellsbank and to be honest I hated it there. We did have good neighbours in Jim and Betty and Flora was an old friend of my mother's, she lived just over the back, but I did not have any real pals. Most of the girls I had grown up and went to school with, like me were married or had moved away. To be sure I also had acquaintances that I met at the mother and toddler groups twice a week but I think it was just something about the place so I can't really tell you the reasons why.

We were finding it hard to make ends meet and the girls did not see much of their dad when he worked in Cumnock. Sometimes they would be in their beds before he came home from work. I could see how tired he was especially when he had to walk or thumb a lift home. So all in all I was happy for Peter to apply and hoped for all our sakes that he got the Job.

It was hard for us waiting until we got word for him to go for the interview, then harder still until we finally knew that he got the job. Then we were all excited and couldn't wait to live in a lighthouse. The only sad thing was that to start with we would not see Peter for weeks on end and at first I was not sure if I could handle the separation. The girls, I think were too young to fully understand or feel anything bad about missing their father. For me it was different, this would be the first time that we had been separated since we were married so I missed him a lot.

It felt a bit like the time when we were courting. I remember when I lived with my Mum and Dad in Burnton having to go down to the local phone box at a time pre-arranged in our letters, waiting for Peter to call from Germany. Sometimes it would be raining and I would get wet and the phone box windows would all steam up before the telephone rang with Peter on the other end. I liked it when the phone box was all steamy. It became my own private place and no one could see me inside talking to a man I barely knew but he was my man never the less and it seemed to make our conversations all the more private.

What helped me most in coping while Peter was away were my Uncle Tam and Aunt Jessie, I would go to their house in Patna once a week and Peter would phone me there. But more importantly they were always on hand if ever we needed anything.

There has always been a special bond between us since I was a child and although I loved my mum and dad, Jessie and Tam had a special place in my heart. They treated me as if I were their daughter and I seemed to grow up with Tommy their adopted son more like a brother than cousin. While we were living in Basildon, Tam and Jessie came down to visit us and were upset that the house was so cold. At the time we had under floor central heating that was very expensive to run, so we managed with just portable heaters. We only had Karen then and she used to kick off her covers even when the room seemed so cold. Tam and Jessie left us £200 pounds just so as we could heat the house until the warmer weather came.

The girls used to love going to Patna and it was not long before I could see Karen developing that special kind of relationship and it was like looking back at me all those years ago. They loved it when their dad came home; even more so when he would surprise them with presents.

We got used to a kind of routine when Peter was away and that helped us to cope, we now had plenty of money in the bank so we had no money problems and that was a big relief after the struggles we had during the first couple of years. All in all the beginnings even with the separation were worth it. What matters most is that we all became closer because of it and made the most of the times when we were together.

I was never as happy as when Peter asked if I wanted to be with him at Holburn Head; it was just a pity that the girls could not come too. I also had a feeling that it would not be long before the whole family would be living at a Lighthouse and that Peter would soon be appointed. Call it women's intuition if you like but I felt it very strongly and hoped that I was right.

The hardest thing for me to do was to get on those trains by myself. I hate travelling at the best of times and this would be the first time that I would have made any long journey by myself and that did worry me a little, especially when I had to change trains so often. The worst part of the journey was the train from Inverness to Thurso; I had to stand for much of the four-hour journey. The train was full of Hikers and their backpacks were everywhere making the whole journey uncomfortable. When I was able to get a seat a rucksack fell from the rack above and hit me on the head and I had to suffer a headache for the rest of the journey and did not recover fully for a couple of days.

It was nice to be with Peter and the Lighthouse was lovely, the only problem was that the bothy only had bunk beds so if we wanted a really peaceful night's sleep we would have to spend our nights separated. Most nights we spent huddled up closer than at any other time in our marriage and it was great.

It was like the honeymoon we never had, not that I ever missed not having a honeymoon. At the time we just couldn't afford it or the time off work, I was happy, just being married. Everything around was new to me so my thoughts were occupied just getting used to things. I never expected Peter to be my Knight in shining armour who would whisk me off to some fairytale castle; but in a way our whole being together was like a modern day fairytale. When I tell people how we met they think I am making it up, even more so when I tell them that I accepted Peter's proposal without ever meeting him. So in a way he was my Knight but it was not to a castle that he whisked me off to but his family's council house in Basildon. It was difficult at first for me because at home I was the oldest sister so had a dominant role, here I felt a bit awkward not really knowing my place at all. Everyone had their part to play in the family and Megan; Peter's elder sister had the role of Mother superior. I don't mean that to be disrespectful in any way, more like she kept the peace and passed final judgment when there were sisterly squabbles. I was treated no differently, in a way I was happy because it meant that I had been accepted into the family. There was not much of an age difference between us; I was a year younger than Jan, a year older than Sue and three years younger than Megan. They all had their own circle of friends, Megan was married and lived not too far away, Sue and Jan had their friends and I had Peter, so I did not have more of a relationship with any of them and in that way I did feel isolated. Sometimes it takes me a while to make firm friendships.

Peter's Dad was a funny strange kind of man, he was even more extrovert than Peter but easy to get on with. Before I found a job he used to take me shopping with him; on a Tuesday we would go to Basildon Market, not just to browse the stalls but to buy fresh fruit and veg. There were two stalls in the market and he would go to the stall that was offering the best deals or sometimes buy the cheapest from each. I remember one time when the stall we were at was busy and we were waiting in the queue, when he suddenly said to me "watch this". He then began to look up at the sky, I of course looked up as well and before long nearly everyone around us was doing the same; wondering what on earth was holding our attention. In the meantime the stall holder was calling "next please" and Pop jumped the queue. Sometimes I think he went out of his way to affront me just to see me blush. We were in the Co-op one time; and we were at the counter. Pop noticed that the young girl behind the counter wore a badge with her name on the right breast of her shirt. Pop came straight out and asked her, "What's the name of the other one?" The girl must have been a little younger than me and soon her cheeks had the same kind of blush as mine, we both looked at each other and I could almost feel her glance saying "Is he with you?"

It was about four months before the Council offered us our own one bed roomed flat, it was nice to have our own place and for me to be able to put my own stamp on things. Peter and I have similar tastes or maybe he just lets me have my way but we seldom if ever argue over choices in decor. Nice though having our own place was; I did sometimes miss having the company of family; that would be short lived though because I fell pregnant within the first weeks of our marriage. We had both taken the decision to allow God and Nature take control when starting a family and we were over the moon when I found out that I was pregnant. We were offered a house a couple of weeks before Karen was born. We had named the baby Karen months before as a kind of sixth sense told us it would be a girl, it was a sense that would stay with us for our following children too. We had the telephone installed on Christmas Eve, which was a blessing because at two o'clock on Christmas Morning I went into labour.

The weather was really good to us at Holburn Head and when we were not wandering down to the harbour or going into Thurso we would soak up the sun. I got on well enough with the wives of the Keepers and we all went in a Taxi once a week for a shopping trip into Thurso. They would tell me what it was like being a Keepers wife and the things I might expect. I did not listen to them with any great interest just in case it dampened my excitement or enthusiasm. I wanted nothing to spoil the high that I was on at the moment. All good things must come to an end and I was sad to leave Peter and Holburn Head. Maybe one day we might be sent up there again, the only thing that would put me off would be the travelling and I was worrying about my journey home, thankfully, apart from a long delay at Dunblane it was uneventful.

Something prevented me from making my usual trip to Patna for Peter's weekly call, so it was Tam that brought me the news that Peter had heard from the Board that he was to be appointed to InchKeith. At that moment the sheer relief and joy of the news had me in tears. Tam hugged me not quite sure what the matter could be, wondering if my tears were of joy or sadness, he could only ask "What's the tears for, don't you want to go"? My reply was to hug him a little closer and say that I was just being a bit silly. Being held by Tam made me realise how much I would miss him and Jessie so now my tears were mixed with sadness too. Worst thing was the girls not sure what was happening came to join us and we were all bubbling, except Tam of course, but I thought I saw his eyes glisten a bit.

Tam took us back to Patna with him because whatever stopped us from making the trip in the first place had lost its importance. I just wanted to speak to Peter and let him know how happy I was.

Chapter IX
The Move to Edinburgh & The Relief

There was no pomp or ceremony on my appointment to Assistant Lightkeeper neither was I expecting one, just the letter initialled by The General Manager and signed by The Secretary to the Board. What this did mean of course was that I had seniority and new responsibilities, when the Principal Keeper was off the Station then I was in charge so the buck stopped with me.

It was not unusual for a newly appointed Keeper to have such responsibilities on their first appointment as Assistant; after all I had spent the last 6 months learning my trade so to speak. I would never be so complacent as to think I knew everything there was to know about Lightkeeping but the Board held the confidence that I was intelligent enough to make the necessary judgments when the need arose.

The Board was kind enough to allow us time enough to move our belongings from Bellsbank to Salvesen Crescent in Edinburgh and time enough to settle in to our new home before I had to leave for the Rock.

The removal van was packed with all our belongings, somewhat meagre when you saw the empty space still left at the back but nevertheless it was all we had accumulated in the four years of our marriage. I was proud to note that what we owned now would not fit into the Luton van that brought our things up from England.

I had given the driver instructions on how to get there, it felt more like teaching my Grandmother how to suck eggs, for he knew the area very well and it was more likely that I was the one to get lost. I told him that we would have to make our way by bus and train but first we must give the house a tidy then take the keys to the housing office. Thinking for a while the driver said that was OK they would go back to depot and have their lunch before setting off for Edinburgh.

I had a good idea of what to expect when we got to the Crescent and knew that we were approaching our destination when the taxi turned into a small estate of snow-white cottages with red roofs and light blue painted windows and doors. I turned to the girls and said, “We’re here girls.” Their eyes lit up in anticipation. As we turned fully into the Crescent, the sight of the removal van assured us that we had arrived, but the girls were less than happy at not seeing the Lighthouse as they expected.

Turning a bend in the Crescent it was obvious the four dwellings occupying this end of the Crescent were different; apart from being much larger they were also two storeys and not unlike the house that we had just left in Bellsbank. This type of housing had the title of Quarter Villa, so we knew that we would have neighbours above as well as next-door. There were two blocks one side of the road and two the other. Each block housed Keepers from the four Rock Stations in the Firth of Forth. The first block on the left was for Keepers from the Bell Rock, on the right the Keepers from the Isle of May, next to the Bell Rock was our Block InchKeith and our house was number 6 the lower and nearest the Bell's, the block opposite was for the Keepers from the Bass Rock.

The front door was opened ready for us, and the driver and his mate were sitting on the tailgate drinking tea and eating what appeared to be a cake of some kind, it was apparent from the sight of the almost empty space beyond that they had already made a start at unloading, as they were drinking from mugs and the cake had been served to them on side plates, it did not take long for us to work out that the neighbours who were observing the spectacle were of the benevolent kind. It would be the children though who would be the icebreakers, as like children the world over their inquisitiveness would always overrule what little social etiquette they had gleaned from the parents. Karen and Kirsty were no exception and they were as eager to meet up with the kids especially as they seemed to be around the same age. Margaret and I were only too glad to have their attention drawn elsewhere, giving us time to get the rest of our stuff out of the van and into the house. We could think about socializing once we had the bedrooms and beds sorted, everything else could be done at our leisure.

There was a small square just beyond the front door, and a glass panelled door led into the hallway, the square was not big enough to call a vestibule and one would wonder at its purpose till you realised that the inner glass panelled door offered secondary protection of modesty to those careless or forgetful enough not to close the bathroom door which was in full view now that both doors were open. To the left in the hallway there was a kind of alcove, which was formed by the external stairs of the house above. The hallway turned right ninety degrees and to the left and right were doors that led into bedrooms, one at the front and one at the back. I put the suitcases down in the front bedroom. We knew that we would be coming to a house that was part furnished but we did not know to what extent. The front bedroom had a double wardrobe a chest of drawers and a double bed with a new mattress. All of the furniture in the house had seen good use and seemed to come from the forties or fifties; nevertheless it was serviceable and complimented what possessions we had.

Hanging up in the wardrobe, wrapped in clear cellophane, was my uniform, its crisp stay-bright buttons depicting the lighthouse ran in two rows of three shone brightly from beneath the film. On the floor of the wardrobe were my shoes and my hat. The crown was snow white and made of PVC making it waterproof, at the front just above the peak was the badge; a brass lighthouse in its centre with a braid rope surrounding it.

There was also a cardboard box sitting on the floor that contained two new white shirts, two black ties, two pairs of black socks, two navy lambs-wool pullovers, two thick navy pullovers and two navy boiler suits. Seeing my uniform for the first time I was taken back to my army days where again everything was issued in twos, only this time I had time to appreciate the quality instead of standing in line at the Quartermaster's store waiting to be issued each item of kit as quickly and matter-of-factly as the occasion demanded and having to sign for the said kit without proper inspection. If the shirts or whatever didn't fit then you'd either have to grow into them or suffer the discomfort until time or the army allowed you to rectify the matter.

Margaret reminding me we had other things to do, and the knock at the door brought me back from my revelry

Bob Duthie introduced himself as not only being my neighbour but also my Principal Keeper. I could not believe him to be the same man that Colin had described at Holburn Head. Far from having an imposing demeanour, I immediately found him to be more of a benevolent father figure and his soft-spoken voice belied the warmth and sincerity of his welcome. Margaret joined me at the door, we could now complete the introductions, though we could hardly make Bob welcome in our customary style given the present circumstances. Margaret attuned to the fact was more than a little shy and apologetic in her response to Bob's pleasantry comments.

Apart from the introductions Bob's mission was to get me to go round with him and check off the inventory of the furniture the Board had supplied and sign for it as well as sign for the uniform, he also wanted to show me my shed at the side of the house, where he had cobbled together some stores. Every year the Board would send out stores such as buckets, mops, basins, and various other household items. Included in the stores were tins of white gloss paint and emulsion and paint brushes. As this was now September and the annual stores would normally be distributed in June. Bob could only manage to cobble together what he could, nevertheless it still appeared as though Christmas had arrived and it was something Margaret and I had not expected, just as we did not expect to see new duvets, sheets and covers laying neatly folded on the beds in each room and the new set of pots and pans as well as a couple of well used ones in one of the kitchen cupboards and crockery in another. It was the sudden and unexpected that made Margaret's eyes start to well up and for her to make her excuses while she went to the toilet to compose herself.

Bob's other motive was to invite the family round for tea just as soon as we could comfortably afford the time to do so. This was the icing on the cake and we knew that we had truly arrived into the family and fraternity of Lighthouse keepers.

Apart from getting the house more or less to our liking, the Board had kindly given us a wallpaper allowance to decorate some of the rooms in the house; we would spend the next three weeks till I went to the rock; settling in and getting used to the Crescent and Edinburgh.

It would not be all play while we were ashore and Bob would still have to fill in shore returns and monthly letters as well as pay for the victualling of the rock; in this respect

I would need to go with him to the Bank to have my name included on the account to draw moneys to pay the bills when the need arose. He also took me round to the butchers and grocers to introduce me to the proprietors, so that I had a face to put to the voice when I phoned in the orders out at the rock. That would be one of my duties as First Assistant Lightkeeper on InchKeith.

The Rock victualling allowance was paid quarterly into the Bank and our suppliers were happy to have their bills paid quarterly too; but we always insured that there was enough money in the account to pay them just in case the arrangement needed to change. It would always be the job of whoever was ashore at the time to pay the bills.

I was obliged to cut the grass at my end of the block both front and back, just as my neighbour upstairs and opposite number would do while I was out at the rock. Bob would do the same for his opposite number while he was out at the rock. It was the same not just here at the Crescent but for all shore stations in the service.

Living in the Crescent at the time of our arrival were: Bob and Chrissie Duthie, they had two grown up children. Above them were Alec and Anne Dorricot, they had a teenage daughter living at home and their eldest son Alec was also a keeper in the service who we would get to meet later. Above us was my opposite number, Dougie Morrison and his wife Wilma and their children Kerry, Kylie and Arran.

In the Bell Rock block, our neighbours on the ground floor were The Wyllie's. Just as we were new arrivals at the Crescent the Wyllie's' were just getting ready to leave, so we barely had time to know them. John and Margaret Bambrick with their two daughters Marion and Karen were to replace them a few weeks later. Above them were, Bob and Mrs Watt and their family Robert, Morag and Marion. One of the Principals for the Bell Rock was Jimmy Burns and he lived with his wife Jan in the bottom house, they had 3 grown up girls. The other Principal keeper lived in Oban so we never got to meet him. Above Jimmy was another Jimmy only his surname was Dickson and he lived with his wife Mary and their two sons Jimmy and Phillip.

Across the road from the Bell Rock block was the Isle of May block; living in the bottom house were Davy and Kathy Leslie they had a grown up son and daughter. Above them were Ronnie and Jeanie Gould with their son Ronnie and daughters Shirley and the twins Fiona and Alison. Ronnie's neighbours were John and Carol Douglas with their two boys, Brian and Steven. Below them were Jimmy and Mary Lyons.

Last but not least in the Bass Rock block opposite us, were Donnie and Anne McIver with Colin and Catherine. Above them were John and May Payne with their son John. The two Principals at the end of the block to replicate our block were Alfie and Mrs. Gunn; their grown up children were, Ellis, Margaret and Neil. Margaret was married to Alec Dorricot Jnr. Dougie and Mrs. McAffer completed the community of Keepers in the Crescent.

Whether by intention or luck the Board had put keepers in the Crescent with families of an age range where no child would feel isolated and not have a playmate or friend of the same age or gender. Already our two girls had settled and bonded with the girls upstairs and Anne McIver's two children. We would have no worries about any of the older children in the Crescent because the moment they got home from school they would play together, the oldest ones keeping the younger ones from harm. It really was delightful to watch them all and reassuring to know that you could leave your precious ones in such careful hands.

Every so often someone would leave the Crescent to be replaced by another family; with the Assistant Keepers in the Crescent it tended to be families because not every lighthouse was suitable for children of school age and as there were twice as many Assistants as Principals these stations were at a premium. We were a service built on seniority; to be appointed Principal you sometimes were filling dead men's shoes. The paradox in the service was that the young Principal Keepers generally came from families of generations of Lightkeepers and that explained the comparative youthful looking Jimmy Budge at Ruvaal. If not for the retirement age of Lightkeepers being set at sixty, I could never imagine myself gaining the lofty position of Principal before automation would make it nigh on impossible.

The only keepers to leave the Crescent before it was my turn to leave were; Alfie Gunn, he was replaced by Kenny Weir and his family. Duncan Leslie and his family replaced Alec Dorricot and Ross McGilchrist replaced Jimmy Dickson. Last to leave the Crescent before our turn were Bob and Chrissie Duthie.

It was soon time for me to go out to go out to the Rock for the first time and as this was the first time, Margaret had the company of Jessie, Tam, Jennifer and Tommy who had come up from Patna for a couple of weeks during the half term break. I did not feel the need of reassurance because I had never seen Margaret so happy and she seemed to have made new friends but it was nice to see them and spend a few days with them before I went out to the Rock.

On the morning of the relief's you could you could feel a sense of expectancy in the Crescent. There would be keepers getting ready to leave for their respective Rock Stations and there would be the wives and families of those keepers out on the Rocks eager to see us leave in order to welcome their loved ones home.

I was probably the odd one out among those keepers getting ready to go out this morning because I was actually looking forward to it. I would be forgiven and assured that the novelty would soon wear off and I would soon like them relish every moment I had ashore.

I was left a couple of wooden boxes by my predecessor that although not quite watertight, was made in such a way as to protect my belongings from all but direct immersion in the briny. I'd packed the boxes the night before with everything I thought I'd need for the month.

The uniform fitted well, apart from the shirt collar whose newness always seemed to chaff my neck until washing and starching made it more supple and comfortable to wear. Nevertheless, there was nothing showier than a crisp new shirt to go with the crisp new uniform.

I took my boxes out to the kerbside to await a van that would collect all the boxes and take them down to the docks. Bob had already placed his as had most of the other keepers. Donny and I were the last to do so and I cheerfully bade him good morning. “What’s the matter?” “You and Margaret fall out?” he replied. He had a point, maybe I shouldn’t look quite so cheerful; and just in case Margaret had been observing, I gave her an extra reassuring hug.

It was not long before the van arrived followed shortly after by a black limousine. Apart from Bob, Donny and I, Bob Watt, Jimmy Burns and Alfie Gunn were also going out. The six of us climbed into the limousine. Jimmy Burns stood out from the rest of us but looking equally as smart in his navy beret and battledress uniform. I’m not sure if his uniform made him unique in the service but it not only showed his individuality and reverence to times gone by; but also the skill of the tailor to resurrect a style that had not been in general use since the days of conscription.

The Lighthouse Depot and berth of the Pharos was in Leith about four miles from the Crescent so a relatively short distance by car; it was not long before we were pulling up at the barrier to the entrance of the docks. The Depot was a short distance from the berth of two navy Minesweepers and as we crossed the path of two mate lots their hands came smartly up in salute; mistaking our limousine and its occupants to be high ranking naval officers. We all smiled at the faux pas and their embarrassment as the car turned left toward our depot and not right to the naval base where they’d anticipated.

There would be no briefing or health and safety film at the depot before the Pharos sailed nor was there any hanging about once we were safely aboard. The Pharos had to be within the Lock Gates at a certain time or she would lose her slot, a bit like an aircraft waiting to taxi for takeoff. Leith docks had the advantage of not being tidal, but you had to arrange departures with the Port Authority to get through the Lock.

The Pharos was larger than her two sister ships and as she was the flagship of the service she had facilities to cater for the Commissioners on their Annual Voyage.
She had a large derrick on her foredeck but lacked the helicopter deck at the stern.
Below decks she was much the same as the Fingal but had extra cabins and a larger wardroom for the Commissioners, neither of which I had chance to see inside but my expectations were that they would be fitting for both their rank and status.

The Keepers had their own wardroom if you’d like to call it that; it had part wood part Formica walls and bench seats either side of a table a bit like a third class compartment of a railway carriage, although we were not made to feel as though we so lowly; it was more for practicality especially as Keepers were generally on board for less than a day. For our time on board we were allocated our own steward who as soon as he could, brought us coffee and tea. The keepers of the Bell Rock who would be the last on board would have their lunch made for them.

Bob and I were first off the Pharos, so made our way up on deck just as the Pharos left the dock gates behind and passed the first of the buoys marking the channel. InchKeith was looming ever closer and it seemed no time at all until we had crossed the stretch of the Firth of Forth between Leith and InchKeith and were dropping anchor.

Once more only this time with Bob, I would make the treacherous decent down the Jacob's Ladder into a launch. Luckily for us it was relatively calm and we both managed to find our way safely in the boat and not in the oggin.

A Sea wall protected the landing at InchKeith and we found relief from any swell when we rounded its lee side. It was not quite low water but never the less it was still a climb up the ladder before we could plant our feet on solid terra firma. Bob and Alec were formally making the relief while Dougie and I landed the provisions and gear on the rock. A hand winch crane on the jetty could lift weights up to two tons but the only stores likely to come anywhere near that weight in a single load were the forty five gallon drums of lube oil, paraffin and diesel that were delivered periodically and the twice yearly delivery of coal for the station. Today we just had our gear and the boxes of provisions. They were laid onto a looped rope strop whose ends were tied securely, gathering the boxes together and forming two rings for the hook of the crane to pass through. The boats crew made sure that the hook and strops were secure enough for us to winch the load up to the landing. Dougie and I took a handle each and made light work of winching up the load and even lighter work of lowering the reciprocal load that was prepared on the landing for the return trip ashore.

Alec and Dougie would not be returning to Shore by the Pharos though; love the service as we did; none of us wanted to delay returning to the bosom of our families any longer than we really had to. Going back to shore on the Pharos would have meant a day's cruise on the Pharos until she completed all of the relief's. Sometimes the weather would get bad enough for the Bell Rock relief to be postponed till the next day and the Keepers who were relieved from the Bass Rock and the Isle of May would have to be landed at Anstruther to catch the train or get a taxi back to Edinburgh. It would not be an isolated case that Keepers arrived home in the wee hours of the morning.

Luckily for us we had an arrangement with the Pilot boats crew that so long as the Pilot did not object and it did not compromise their service they would pick the relieved keepers up from InchKeith just as soon as they had opportunity to do so. Invariably and with their good grace we were more often than not home just about lunchtime, the pilot boats were the fastest craft on the Forth and Granton was their base, just under two miles from the shore station.

As the Island was new to me I was eager to get started, so I did not envy Dougie going ashore. Envy would come much later once the novelty wore off.

From the landing you could not see much of the Station because all of the lighthouse buildings were obscured by the elevation above the landing. Only the top of the dome was visible. A snow-white wall trailed in an inverted "L" gave some indication to the boundary of the Station. The tell tale outlines of a roadway with its hairpin bends zigzagged down from the Station to the landing. At the base of the hill facing toward the road was a Massey Ferguson tractor and trailer. I did not need much encouragement to begin loading our gear and provisions onto the trailer. The tractor's engine was still, but the chug, chug, chug of an engine could be heard; tracing the sound; I detected a squat brick building some fifty feet up the first gradient; painted the traditional snow white and green. I had no doubt the Bob would tell me all I needed to know about this and other things once he had finished talking to Alec and the relief made. Since the advent of the telephone most of the details of the state of the station were passed from Principal to Principal the night before the relief. We were unusual in the respect that reliefs on our Rock were so casual and not done in the haste that was my experience so far. Even so the Rock Station could not function for too long with only one keeper at the helm so to speak, so Bob and I made our farewells to Alec and Dougie and made our way to the tractor.

Bob was now free to tell me more about what I could see on this side of the island and started with the answer to the engine noise especially as painted in lighthouse colours it obviously had something to do with the station and therefore formed part of our duties. Once we had made our way to the structure; I could see that it was built with bricks on three sides and had the rock itself as its back wall. The door was opened and I could now see in the dimness of the interior the single cylinder Lister engine sitting squarely in the front half of the room. Behind the engine was a deep square trough whose sides went below floor level; water trickled constantly into the trough from a crack in the rock surface just above the rim of the back edge. "That is the source of our water" Bob said; he continued, "It comes all the way from the Northern half of the Kingdom of Fife." He briefly went on to tell me the history of the well and how, long before there was a Lighthouse, ships would use the well as a source of water for re-provisioning rather than use the ships chandlers in Leith whose service came at a costlier price. There was a price to pay for the continual use of this wells water because it collected calcium on its journey. The sides of the well were heavily coated in the chalk like substance, so much so that I naturally assumed that it was lime-washed or painted annually. Bob advised me not drink the water without boiling it first, not that it was not pure or clear but just so that the lime-scale would more readily separate and line the kettle rather than the internal workings of our bodies. There was a plant the station for extracting the calcium but it was waiting for parts to repair it. In the meantime, time served on InchKeith was reduced as much as possible and Keepers seldom were at InchKeith longer than three years, given our time ashore, exposure would be less than 18 months. No one had reported having trouble with gall or kidney stones throughout the history of InchKeith but never the less the Board tried everything to reduce the risk.

It was an easy enough task to close down the engine of the water pump; just draw across a lever and pall which prevented combustion in the cylinder, to start the pump all one needed to do was to make sure there was enough fuel in the tank and provide the necessary impetus on the starting handle, once achieved push across the lever and the cylinder head would begin compression and combustion.

This was a relatively easy task while the weather was temperate, but in the depths of winter when your hands were freezing and the engine was in a belligerent mood then there would more than a few curses coming from the direction of the well; given a favourable wind and the sharp ears of a lighthouse keeper with nothing better to do.

There were the remnants of other buildings to the right of the station and it was obvious that at one time in its history Keepers were not the islands sole inhabitants.
I do remember my father telling me that while he was serving in the Army, he would go out to the islands of the Firth of Forth and service the ammunition for the anti aircraft batteries and the larger guns especially those on InchKeith. I did not think that there would be so much if anything left of their presence. I could clearly see the foundations of the wooden barrack blocks as there was one in near complete form still standing as template for the others. There seemed to be a more permanent structure of brick; which was undoubtedly the officer's quarters and mess. Further up the hill you could just make out the entrance to the as yet unseen fortifications. An overgrown roadway led from the right fork in the road ahead of us and passed a large shed that I thought might house a boat of some kind. Then the road snaked its way up the hill on the opposite side of the bay from the jetty servicing all of the buildings at the different levels.

I could not wait to get exploring the island but for the moment we had other more pressing things to do. We locked the door to the well and made our way back down to the tractor. Bob had asked me if I could drive a tractor and was assured by my reply, thing I never told him was that the last time I drove a tractor I got a strong rebuke from my employer for breaking three brand new tines from his harrows and at this moment Bob's confidence in my abilities far outweighed my own.
I depressed the clutch and selected second gear, I knew that some tractors had a solenoid inhibitor in the clutch that would prevent the engine from starting unless the clutch was engaged, so I kept my left foot on the clutch then tuned the ignition key. The engine responded at once and saved me a moment's embarrassment; I gently released the pressure on the clutch and felt the bite. I released the handbrake and at the same time released the remaining pressure on the clutch smoothly and gently the tractor crept forward without throwing either I or my passenger forward, all I had to do now was to apply enough acceleration to take us up the hill without losing control on the hairpins. "Come on Peter, it's not the Monty Carlo Rally," I kept telling myself. But all the while at the back of my mind I could hear the twang as the tines broke as the harrow caught on the fence as I turned through a gateway from one field to another.

This time my driving was faultless as we climbed the last rise and entered the courtyard. There were buildings on the right in a state of disrepair that were far enough away from the Army dwellings and separate from the lighthouse buildings to suggest that these were to serve a purpose independent from both. As it turned out InchKeith was once a Lloyd's Signal station and these buildings were to house the operators. Behind them was an old octagonal wooden structure that was used to test experimental lights, well at least the top part was octagonal, the middle part was wooden but twelve sided and the base was round and made of brick. The whole structure was about twenty-five feet tall; bereft of glass and open to the elements making it completely unsafe. The Lloyd's buildings were not much better so I would avoid going into them on the strong advice given by Bob. I could tell by his intonation that it was not merely advice but an order; as well as respect I have a good deal of common sense and knew there was nothing in these buildings worth investigating.

The Station could now be seen in its entirety; a shed was to the right of the courtyard entrance housed the trailer. The tractor was housed in a shed to the right of the Tower Block. To the left was the boundary wall and ahead of us the wall stopped at what looked very much like a second world war pill box or look out post. My second judgment was nearer the mark because the Lloyd's operators used it for just that purpose. Because it formed part of our boundary wall, we the Keepers would paint and maintain the structure just as if were an integral part of the station. The wall continued on but at a much higher level until it formed part of the Lighthouse Tower building; unusually it was not the customary snow white but a brownish sandstone colour except for the windows and doors. What was also obvious was that the building was designed to mimic a castle as it had ornate parapets on the roof facings of its square walls and the tower itself looked like a turret protruding from the centre. All of the windows were bereft of curtains and lacked the signs of occupation, but nevertheless you could sense the presence of former occupation and imagine the ghosts of keepers past, when this was a family station and the island was a bustling community of keepers, Lloyd's operators and Army personnel.

Bob directed me to make a "U" turn and pull up outside the familiar shape of the engine room block that occupied the remaining side of the courtyard; there was a wall that joined the two buildings together with an archway. Above the arch was crest with a coat of arms with the date 1556. It looked as if the keepers had tried their best to maintain the crest but time and paint had caused the relief to be somewhat faded and it was only just legible. I thought Bob was joking when he said that Mary Queen of Scots was a frequent visitor to the island, but then my knowledge of Scotland's history was in its infancy; this revelation would act as a stoker to reduce my ignorance. Through the archway I could see the familiar shape of the receivers for the foghorn so knew what to expect beyond the engine room doors.

Tagged onto the end of the engine room block was the living quarters, I could see the curtains hanging in the windows. Protruding out from the block was the entrance vestibule and standing in the doorway was John McInally one of our two Local Assistants. He would be relieved in a two weeks time by Jimmy Coombs. For now he would help me unload the trailer and hopefully make the customary coffee as the time was approaching eleven o'clock.

My room was at the front and overlooked the courtyard, it seemed overly large for a bedroom and made larger by the absence of the furniture to fill it. A single bed, a chest of drawers, a wardrobe and a small table and chair made up the contents. My first impressions were that I was glad to see a radiator under the window and even more relieved to feel it's reassuring warmth when I touched it. It was the end of September and already the sun was beginning to lose the balmy heat typical of our Scottish summers, I reckoned that I might need more than a duvet to keep me warm as winter progressed and was happy to see that there were additional blankets on the top shelf of the wardrobe. My unpacking was cut short by the call of "Coffee up," coming from the back of the block where the living room and kitchen were.

This would always be the time for formal introductions but also the time for Bob to go over any schedules for work or intimations of visitors to the station, thankfully there were no visiting personnel due and the only work that was planned was the recovery of some timber from the old army buildings, but that would only commence once Bob was sure that I was familiar and competent to do my duties as 1st Assistant.

InchKeith had some features of lightkeeping that I was unfamiliar with and needed instruction on and after coffee break and once I had put the tractor and trailer away Bob would begin my tuition.

Chapter X
InchKeith
The Station, The Island

InchKeith was one of the few remaining Lights that were lit by a paraffin incandescent burner; this was the closest to the lightkeeping of a hundred years ago. The light produced and enhanced by the lenses was still around a million candelas; more than enough to carry the distance across the Forth from shore to shore and adequate to warn ships sailing up the Forth from the rail and road bridges to the west; to the Isle of May in the east.

To compliment the light; the station also had a Radio Beacon sending out its signature in Morse code to the radio operators and navigators on ships; tuned into the frequency in order to help them get accurate positioning especially in thick fog. It had its transponder in the tower but you could monitor the signal by tuning the HF radio or for that matter any radio, the signal was strong enough at close range to bleed over a few frequencies. To get to the radio room you went through the kitchen and pantry and through a doorway; there were a couple of radio units sitting on a bench that occupied the right side of this narrow room. At the far end of the bench sat a taller unit that had a variety of switches on its face. Aerials came out of the back of all three units and the only unit I could identify for sure was the familiar HF set that was in use at other stations. As it turns out the antiquated looking VHF set was no longer in use and the tall unit was the monitor for the fog signal out on InchColm. The fog signal or emitter there had sensors that would automatically sound the emitter during fog, all we needed to do was to switch the monitor on and listen to the sound coming through the speaker. If we could not hear it sounding then we could start the signal remotely by pressing a few switches, if it still did not sound we would have to report the fact to the marine superintendent. It was also possible to use the monitor as a VHF transceiver as it was tuned into the lighthouse frequency on another band and be able to speak to the other rock stations. This was done especially at Christmas; you could hear the chatter between the keepers and know that the reception was good.

I made a bit of a fool of myself the first time I tried to make contact with the other rocks and the fact that the mistake was so elementary had Bob in fits of laughter to my further embarrassment he would refer to the incident every time I was in danger of becoming stroppy. It was Christmas morning at about ten o'clock and the monitor was on to hear any banter. I could hear Donnie McIver talking to Jimmy Burns and was hoping to bid them both seasonal greetings. Ignoring the send and receive switch that I had used enumerable times when monitoring InchColm; I chose to switch on the remote switch in the vain hope that I could send and receive at the same time. Several times I called with no response, in the end I gave up. Bob asked me later if I had given the rocks our compliments of the season and was perplexed to hear the lack of response. He straightway went to the monitor and switched it on, voices of other keepers could be heard; waiting for the appropriate moment to transmit' he pressed the send switch and began his conversation; the response was almost immediate as was the reddening of my cheeks. It was obvious to Bob that I had recognised my mistake but it would be bad manners for him to make enquiries until he had passed some pleasantries with the other keepers. He did not let it pass and was intrigued to find out what I had done or hadn't done. His laughter began just as soon as I had finished telling him. It just so happened that at the time of the incident; one of the most popular TV adverts was the one by the RBS depicting the man standing outside the bank in front of the auto-teller asking the machine, "Is the Manager in?" I must admit the comparisons were pretty remarkable and I even found it funny but the humour wore a little thin, especially when Bob in order to win an argument would remind me of the incident with the repetition of the catch phrase.

Bob was a great teacher and had a wealth of knowledge that I found unsurpassed in the service, there was not a great deal that Bob did not know about Lighthouse Keeping, he had exacting standards that I could understand Colin McEwan mistaking for over zealousness but given the nature of the light here on InchKeith it was only as good as the maintenance that was rigorously performed on it. To that end it was the responsibility of the 1st Assistant to clean and maintain the paraffin lamp.

In the summer months the light generally only needed cleaning once a week but the longer the light burned the more carbon collected in the parts forming the lamp and it

required cleaning twice or sometimes three times a week especially towards the winter solstice when the nights were longest. I had never been to paraffin light before so it was important for me to get used to the lamp and all its parts and to ensure they were cleaned and free of carbon deposits.
Bob knew that I had served in the army so all he needed to say to me was that if I cleaned the lamp as well as I cleaned my rifle then all would be well. Taking good care of my rifle could be paramount to saving my life; taking good care of the lamp was paramount in saving my job so I looked to the task not with trepidation but with pride.

To this end Bob's first priority was to take me up to the Lightroom and bring down the lamp for cleaning, his thinking was although the lamp should always be in a condition of readiness for use; the relief changeover of personnel could not be used as an excuse should the lamp fail on the night of the relief, there was only one way of resolving the matter and that was to clean the lamp. So we left John to his kitchen duties and crossed the courtyard to the tower block.

The door to the tower opened up into a small vestibule, ahead was the open staircase leading up to a landing, then upward again to the Lightroom. A dark passageway led off to the right of the stairwell and you could just make out some doors that would presumably take you to the rooms of the main building, we had other things to do at this moment but I would investigate properly when I had the chance.

At the bottom of the stairs sat a large can with a spout, a bit like a watering can. The telltale smell of paraffin and the bung plugging the end of the spout instead of a rose indicated that its use was not horticultural. Bob did not have to say much other than "We'll take this up with us." That was another thing that I liked about Bob, he never came out with a direct order or request, he always used the term we will do this or we will do that, knowing perfectly well that the task would only require one person to perform it. As he tended to be one step ahead of me in most respects, I would be left in no doubt of his true intentions.

The set up here was much the same as Killantringan and once you got beyond the first landing you could well imagine what lie beyond without being too surprised at the differences. Carrying the full can up the last ladder to the Lightroom was a bit of a struggle and I was only too happy to be able to set it down. Bob remarked noting my efforts, "Wait till you have two cans to carry and the icicles are forming in your nostrils." Although I could imagine it, I naturally thought it was good-humoured exaggeration.

The Lightroom and indeed the Lanternroom were very much like that of Killantringans', the major difference was that instead net curtains there were roller blinds and instead of a mercury vapour lamp there was a paraffin lamp. There was also a pair of cylinders that looked a bit like Oxy-acetylene bottles, only stubbier and painted blue. They were connected together by piping and one of the bottles had a hand pump; a bit like a bicycles in front of it. A pressure gauge on the bottle gave the final clue that one of the tanks contained the paraffin and the other pressurised air. When the valves were opened paraffin would travel up the piping to the lamp above. The tank containing the paraffin had a curved pipe welded to the top and this pipe was capped with a big brass nut. This was the filler neck for the tank.

Bob showed me a spare cap and I could see that it was lined with lead. Every so often the lead would need to be heated up, melted and re-set so that it formed the seal for the tank when it was screwed tightly.

We went up to the Lanternroom to remove the lamp from its stand. Just like Killantringan; the lenses had a section that opened up allowing access to the lamp.

To best describe the lamp it's better to break it down into the three constituent parts; starting at the top where the mantle sat on a brass collar, the delicate gossamer was so fragile that great care was needed when removing the collar from the top of the lamp housing. The lamp housing had metal gauze which was set in a brass collar that screwed into a cylindrical brass chamber the same diameter; it had a pipe that came out of side of the inverted copper funnel like top of the housing welded to the main body. The interior was hollow and open at the base. An inch above the base was an oval opening through each side strengthened by brass plates. This opening was to house the vaporiser. The base of the lamp was finished with a brass collar that would act as a mount for the stand in the Lightroom. The vaporiser was like two hollow tubes of gunmetal clamped together with a gunmetal collar at each end. At one end of the vaporiser the tube was threaded and a copper and brass pipe in the shape of a question mark was fitted, the pipe next to it was capped off at the end with a screw bolt. A fine hole was drilled into the screw bolt at the bottom of the U shaped vapouriser to allow the vapour to escape. At the other end of the vaporiser and the opposite end to the brass pipe, the end was capped off again but a small hole was drilled into the collar linking the two chambers together. Paraffin would travel through the pipe and be heated till it turned into vapour escaping at the other end. When the whole lamp was assembled the vapour would come out of the nipple with force and enter the chamber via the brass pipe at the side of the funnel shaped top of the housing to be burned like a gas through the mantle. It was my job to ensure that the whole assembly was clean and free of carbon that might break free and block the nipple, the chambers or the hole separating the chambers causing the light to fail. This job was performed in the workshop where the necessary tools were kept such as brushes, carborundem paste, prickers' and scrapers

Once the lamp was cleaned to Bob's satisfaction we returned to the Lanternroom to test the lamp and for me to be taught the important art of forming a mantle. Before the lamp could be lit the vaporiser had to be pre-heated; lighting two wicks set in a brass container filled with metholated spirit did this. The container was placed into a mounting that swung beneath the lamp assembly and directly beneath the vaporiser. After about five minutes or so the vaporiser would be hot enough to turn the paraffin to vapour. While it was heating Bob taught me how to set a new mantle.

The mantle came in a packet and was like a deflated balloon made of silk. The open end had a band of pink silk and woven into it was a drawstring. The mantle was tied onto the brass collar of the mantle holder ensuring that the drawstring sat neatly in the groove made for it. A metal hook slid into a mount at the side of the collar to support the top of the mantle during forming, it would also provide a gripping point to allow you to place the mantle onto the lamp without burning your fingers. There was a special set of pliers made for the job that had grooves to the same diameter of the hook set in it the jaws, keeping your hands even further away from the heat. To finish off the assembly I had to gently tuck the gossamer into the collar. It was not essential to do this but Bob reckoned that it helped in forming the best mantles.

The vaporiser was now hot enough so all I needed to do was open the valves to allow paraffin up to the lamp, after an initial spurt the vapour was clear and went into the chamber. I was instructed to light a taper and then light the top of the lamp without the mantle. There was not room enough for both of us so all his instructions were spoken. All of the tools needed for lighting the lamp were set on a tray beneath the lamp pedestal so it was just a matter of following his instructions, after a few sparks the flame flickered with tinges of red, pink and violet but the basic colour was a vibrant blue. The heat was intense and looking into the flame was not good for the eyes, to prevent serious damage I needed to wear a pair of ultra violet protection goggles that had deep blue lenses. You could hardly see anything out of the glasses except when you looked directly at the lamp, all of the previous colours disappeared into a uniform white. I was instructed to pick the mantle assembly up with the pliers and gently place it over the flame then to gently lower it onto the housing; at the same time I had to gently blow on the mantle. As soon as the assembly was over the lamp the heat began to inflate the mantle just like a balloon, but very slowly. Blowing gently on the mantle and gently turning it on the housing encouraged it to form uniformly and that was where the art was. The more uniform in shape the longer and brighter the mantle would burn. Because the mantle was so fragile even a puff of wind might cause it to collapse. It did not happen on this occasion and both Bob and I were proud of my efforts. “Well done Peter, it’s not often keepers get it as good as that on their first attempt, so don’t be too disappointed if future attempts aren’t quite as good.” My tutorial for the day was over and Bob was reassured that the lamp was clean and functioning well and that I was capable of standing watch. Tonight I would be lighting the lamp for real and be standing my first watch as the 1st Assistant Lightkeeper at InchKeith.

It would take me more than an afternoon to see everything there was to see on the island, even after three years there were things that may have been overlooked and as yet undiscovered, so forgive me if I give the impression that all could be revealed in one excursion, nevertheless I’m sure that most would come away from the island with a sense of fulfilment after such a visit.

Because of the family connection with InchKeith, I felt beholden to search the visitor’s book in the hope that Pop had left his mark. Alas he must have been too busy or eager to return to other duties ashore to have time to visit the Station. I, on the other hand was obliged out of custom to record my details for posterity and of course for my personal desire to leave my mark in case future generations of Hills had the good fortune to visit the island.

The visitor's book sat on the desk in the Lightroom and flicking through the pages I could see that for a rock station, it seemed to be either a popular attraction or as was a more likely case; a hive of activity, in a continual state of flux as personnel servicing the island garrison were exchanged on a regular basis. The earliest signature in the book was dated July 1944, so it could be that my father was here prior to that. It was unlikely that the previous visitor's book was still on the island as it was customary to send the full book to the Secretary. That was the other thing about Pop, you would get the basic story from him but you seldom got the whole thing; so like a jig saw you had to piece the information together one piece at a time in the hope of eventually coming up with the right picture. Unfortunately all I seem to come up with is an abstract, maybe that was how he wanted it.

What I could glean from the book was that the volume of signatures seemed to tail off in the sixties as the interval between signatures increased, so that must have been when the garrison was de-manned and closed, leaving the Keepers and their families as the only inhabitants. At the same time the Station would have become unsuitable for families; with less activity and travel to and from the island; the Board would have the responsibility and the cost of schooling the children.

The evidence that this was a family station was all too clear in the dwellings of the tower block itself, firstly in the Lightroom there was a bell push system for calling the next man on watch and it was marked Principal, 1st Assistant, 2nd Assistant and 3rd Assistant. Had it always been a Rock station there would not have been the need for the third assistant and secondly there would not have been the need for the amount of dwelling space. The tower block had two storeys; the most likely configuration was that the Principal Keepers house occupied one of the floors. A sense of imagination had to be used to see the actual layout of the flat because all of the utilities had long since been removed leaving all of the rooms bare and uniformly bland. It was the same for the ground floor and as a whole, the layout of the station left more questions than answers, as there now appeared to be more Keepers than dwellings. Bob could not help solve the riddle either, as the restructuring of the station took place long before he was here. The overall impression was that the Board had plans for the tower block dwellings but had abandoned the idea. Short of unearthing the wires of the bell push system and tracing their course; which incidentally would not only be a major undertaking but highly frowned upon by the Board, then the mystery would remain in the archives of 84 George Street or for future archaeologists to discover.

My inquisitiveness was rewarded in one respect in that while looking through the cupboards of the ground floor, I was able to find not only the previous visitor's book but also the original Station Bible. Neither could provide the evidence that I was looking for in order to satiate fully my desire for answers but the first did establish that my father did not sign the visitor's book. I need only look as far back as 1940, one of the only dates he did tell us about in reference to his second stint in the army and the date of his call up for duty. He was bitter about being made redundant so to speak during the recession of 1929 so was reluctant to volunteer at the outbreak of World War

II. "If they want me they could damned well call me" he said, knowing full well and anticipating that they would do exactly that.

The Station Bible did offer a suggestion that because it had accumulated a layer of dust, it must have lay there for quite some time. If so it might indicate that this was the Principal Keepers house, as it was traditional for him to act as pastor and perform Sunday service for those religiously inclined. This was more common in the early nineteenth century rather than the latter half of the twentieth. I could just picture the scene of the Principal Keeper standing proud in his naval frock coat delivering a sermon to his gathered flock of Keepers and their families all dressed up in their Sunday finery.

One of the rooms in the lower flat did seem large enough to perform that function.

There was little else of note in the building so I made my way out to see what might be of interest in the lookout post. As there was just the one I could not fully fathom its use as you could only see an arc 180 degrees westward from Leith to the south and Burntisland to the north; all of the mouth of the Forth to the east was obscured by the lighthouse buildings. So the observer could only see shipping on their way out of the Forth and not what was entering. It was made of concrete and devoid of anything that would give anything other than basic shelter; it still remained open to the harsher elements of wind and cold.

It was from the vantage point of the Lanternroom which co-incidentally was the highest sheltered point on the island that an observer could see the full 360 degrees. From there you could also make out the general shape of the island, the nearest would be to describe it as being like a teardrop with the pointed end toward the Leith in the south.

A ridge rose up along its centre rather like a spine that gradually declined toward the tail. The lighthouse was on the highest point of the ridge on the broad end of the island. Beneath the ridge the ground seemed to flatten out into a kind of plateau and then it fell sharply some fifty feet or so into the sea. At the end of the plateau and nearest to Burntisland sat the foghorn.

To the left of the foghorn the ground rose twenty feet or so forming a ridge that was rocky and without vegetation giving it a separate identity; the roadway trailing down to the landing acting like a cleft between the two.

The landing was almost a natural feature as the land had a slight curve to form a shallow bay; the addition of the concrete pier and breakwater provided the extra protection.

The island tapered on both sides and ended in a cluster of large jagged rocks. Green and red navigation buoys marked the south channel between Leith and InchKeith but there was an extra buoy just off the end close to the rocks and this marked the spot of a wreck of a vessel that had ventured to close to the island.

Apart from the shoreline, the centre where there were no structures was lush with vegetation helping to conceal the gun emplacements to all but the closest scrutiny. I was told they were there but there was little evidence to support it and being a former gunner I was more than just a bit interested to see what remained of the batteries.

Bob was on watch so it was incumbent upon me to inform him where I was going; sarcastically he said, "Don't get lost" with the reminder, "Be back in time for tea." As there would be steaks for tea, he could rest assured that being late would be the last thing on my mind; remembering Rabbie Burns and past intentions I decided to make the gun emplacements my only venture for the day.

I decided to walk the top of the ridge down past the Lloyds houses and the old light tower, judging that this route would make it less likely to for me to miss the emplacements. I could not see how I could miss them; after all eight inch gun housings would not be the easiest thing to conceal even without the gun barrels sticking out. However following the ridge it was impossible to see their exact location, the vegetation undulated regularly but was heavy enough to hide the corrugated effect causing me to trip more that once as I stumbled from furrow to furrow. I could also see that the grass was dotted here and there with the remnants of nests. Coming across one nest that still had two eggs in it; I came to the conclusion that there could not be many rats on the island; as they would not have left such tasty morsels as these long enough for them to become addled. I can't remember ever seeing a rat on the island either. Perhaps one of the Celtic Saints did for InchKeith what St Patrick did for Ireland, come to think of it in all my time at InchKeith I only ever saw the last few inches of what I took to be a grass snake as it slithered into deeper grass.

I had remembered this morning seeing the entrances to the emplacements from the landing and mentally noting their position from other landmarks but it was not until I was twenty feet or so away from one entrance that I knew for sure.

Whether the concealment was by design or an accident of nature; taking over after man's departure they were far better hidden than any camouflage used by our regiment. I remember being taken up in an army helicopter that had visited us while we were on exercise. The troop commander seemed quite impressed by our attempts at camouflage and made remarks to the effect, he did however make the mistake of asking my opinion. His bubble had been burst with dramatic effect when I ventured my comment that the camouflage was fine up to the point where the radar antennae of Big Ears and Noddy could not be concealed and therefore one good placement of a napalm bomb could wipe out the whole troop. He did not appreciate my candour and I think it was a few days before he spoke to me again. So I was reminded that camouflage was only as good as the lack of intelligence behind it; because the gun emplacements were sighted on a known island and the island was definitive in both its size and shape; any aerial bombardment would be bound to find the target. Only the reinforced concrete and mounds of earth would protect it from complete destruction and of course the lack of the will of a determined enemy to face the anti aircraft batteries that protected Edinburgh as well as the Firth of Forth. No I think nature was allowed to take its course and made use of the fertile soil covering the surrounding area; the thick tufts of grass would bind the soil together and act as a kind of matting absorbing to some extent the explosive impact of any bombs. There were no craters to be seen anywhere on the island, either they had been filled in or the island never came under direct attack from the air, so it looked very much like the Luftwaffe had more important targets and the Kreigsmarine lacked the will to sail up the Forth to attack Edinburgh from the sea.

The entrance to the Battery was narrow and vault like, the interior dark. I would need to return with a much more powerful torch to see into the furthest recesses of the interior. The light that my simple torch offered was barely enough to guide me through the maze of tunnels that went right and left beyond the entrance. On one route the tunnel ended in a flight of steps that took you deeper into the hillside and entered a vault like room. Light was coming from the ceiling at one end of the room and the traces of an ammunition hoist could be seen, the hoist continued down below floor level to what I imagined must have been the magazine. I could not work out what was kept in here but knew that above would be where the business part of the battery would be mounted. All I had to do was find the right way and the right door to take me there. It took more than one attempt to retrace my steps back to the entrance; reinforcing my idea of bringing a more powerful torch and adding a piece of chalk to the list as an afterthought.

I was not too disappointed in not seeing the eight inch gun still sitting on its mount and besides; the temptation to fire a blank shell in response to the one o'clock gun fired from Edinburgh Castle would have been too much for anyone to resist. The percussion would have resounded to the Castle and beyond, by far out ranging in sound the meagre limits of the twenty-five-pounder. The only remnants in the emplacement were the rusting bolts of the mount brackets, everything salvageable had been removed, and the same could be said for all of the rooms that revealed nothing when I returned for closer examination with a brighter torch. The amazing thing was that in both emplacements the tracking and chains of the ammunition hoist were rusty but still in place. Maybe they were not worth the effort to dismantle or perhaps they were just forgotten and overlooked as being the last pieces to be dismantled. I would come back to the emplacements every so often and always left them with the thought that there was something they had kept hidden from me; a bit like my father I thought well that's maybe how it should be.

I could not find many traces of the islands previous use before the lighthouse was built and remembering Jimmy Budge's story about the island being used as a Prisoner of war camp during the Crimean War, I'd hoped to find some evidence, but could find none. The lighthouse pre-dates the Crimean War by some fifty years and I found it hard to imagine the Keepers acting as Warders or avoiding contact with the inmates on such a small island.

There would be other days and other Keepers with a greater thirst for history who might uncover more about the Island of InchKeith but for now I would just enjoy my walks and be happy with whatever I found on my excursions.

Chapter XI
Winter & Summer work and the Commissioners visit

From September onwards and while the ground was free of frost and snow we would prepare the last of army buildings for demolition so that we could recover the Canadian Pine that was used un both the base and uprights of the structure.

The Board's carpenter thought that the cost of reclaiming the timber was cheaper than purchasing the wood new and for a time that was the case but eventually the source would dry up and the need and cost make the proposition less viable, luckily for us they coincided so it was not so much of a loss to us when it finally came to an end after my first winter on the island.

We were paid by the hour so on good days we could work four may five hours a day, about the same time that we spent painting during the summer months. Bob and I would be the ones to do most of the grafting because the half relief had come and gone and Jimmy Coombs had replaced John as Local Assistant. Jimmy was in his sixties and preferred to stay in the kitchen during his month, because we all shared in the bounty of whatever work there was going at the station; I preferred to be out working. So it suited all of us.

Bob and I would take the tractor and trailer down to the big shed at the landing and then take the track that led up to the last of the barrack buildings that still stood. Most of the roof panels had gone and some of the side panels were removed, before we could pull the building down, we had to remove all of panelling and the roof trusses.

The Canadian Pine was well-seasoned timber but you did not want to expose it for too long to the elements. The idea was to strip the building down to the frame made of the valued and salvageable Canadian Pine as quickly and carefully as we could, all the other wood such as the panelling and roof trusses had no monetary value at all so was stripped and thrown to one side. When there was just the framework left we had the difficult task of sawing through the cross beams to bring the structure down section by section, there I go now using Bob's meaning of the term We. I was the young and agile one climbing ladders and doing all the sawing while the other half of the We cleared a pathway so that the tractor could get in. The trick was to position the tractor in such a way that you would get the maximum amount of traction, the other trick was to put the chains in the right positions to get the maximum amount of stress on the joints without the wood snapping or splintering if you got them both right the section would come down with the merest of groans in protest. On a couple of sections though; it became more like the immoveable object against the irresistible force, neither wanting to give way. Eventually though with some slight adjustments to the chains and a little friendly persuasion from a ten-pound sledgehammer we managed to get the last of the sections down.

The lengths of timber were too long to fit on the trailer so we had to chain four or five lengths together and drag them down to the shed; once they were all down we stacked the lengths in the shed ready for the next part of the salvage operation. It took about four or five days to strip and bring the wood down to the shed, it would take most of the winter to strip the wood of all the nails until it was finally ready for shipment back to the depot.

When the weather was too cold, Bob would teach me some of the finer points of lightkeeping, he was especially fond of rope work; the need to be able to tie specific knots or to be able to splice rope together though not essential would come in very handy indeed when the time came.

The most common knots we used were; half hitches to tether the boats to mooring eyes or hand rails, reef knots to temporarily join two pieces of rope together, barrel hitches to secure barrels together and secure them to the trailer. Rab Kyle, a Supernumerary Keeper, who coincidentally hailed from Ayrshire and was only junior to me by a couple of months; was out with us on two occasions, replacing either John who resigned or Jimmy who retired. He taught us how to do a lorry hitch that was a much better way of securing a load to the trailer. The knot that we hoped we'd never have the occasion to use but nevertheless the knot that could save our lives was the bowline. If ever we found ourselves in the drink and we were lucky enough to be thrown a lifeline, it would be the only knot guaranteed not to slip or loosen.

It's amusing to think on it now, but every time I see the film Jaws, I am reminded of the way that Bob taught me how to tie the knot and wonder if was taught the same way since the its first inception; "The rabbit comes out of its hole, goes round the tree and back into its hole ". I would like to believe that the expression tying the knot in reference to marriage refers to the bowline because of its merits and not to the symbolic bond joining the betrothed couple together in the marriage ceremony.

Splicing rope has some euphemistic terms not found in the dictionary or used in the manual to seaman, the reference book Bob had kindly let me borrow in order that I might practice the more commonly used knots and splices and learn about the more decorative and less practical ones. The back splice or "Dogs Cock" was used to prevent the end of length of rope from fraying, the eye splice was used to make an eye at the end of rope; to make the rope strops all we would do was to splice a length of rope together. The strop was used in two ways; if we were lifting our boxes then we would lay the rope down, stack the boxes onto the rope, gather the two ends together and tie them tightly with a single half hitch. The weight of the load would be kept even and the hook would have two eyes for added strength. If we were lifting barrels then we would lay the barrels down on the strop and slip one end of the rope through the other, the rope would tighten onto the barrels as the weight came on. If we used the same method with the boxes then the extra strain might crush the boxes and the contents inside. There was one other knot that took a little time to master and that was the Monkeys Fist; this knot was designed to form a weighted ball at the end of the rope and the extra weight would allow you to throw a line straighter and further than a coiled rope.

John McInally would look on with mild interest while I was being tutored but never showed the interest when asked directly to join the class. That was the thing that kind of separated John from us, he was a good enough keeper but it seemed as if he treated the service as just another job, where Bob and I saw it as a vocation. It was best expressed when on one occasion Bob and I were discussing the nature of spring and neap tides, I was interested in knowing the difference; in answer Bob was talking about the phases of the moon and John overhearing us chirps in, "What's the moon got to do with it?" From his tone of voice this was more of a statement than a question and all Bob and I could do was to look at each other in puzzlement. Later on I tried to explain to John the moons influence on the tides but he was completely disinterested.

I deeply regret that I was unable to find the common ground to strike a bond with John but maybe that was his choice and he was happy enough just to do his job and go home at the end of his month. I do hope that I did not contribute in any way to his leaving the service because it seemed such as short time after my arrival that he resigned from the service.

Jimmy was a lot easier to get on with and we had the common bond of both seeing service in the army, Jimmy was in the paratroopers and had risen to the rank of RSM.

Jimmy was a good few inches shorter than me but that would not be an impediment to reaching such a lofty rank, after all my own RSM at 17th training Regiment was even shorter. I found it hard to keep a straight face when being admonished by him; I could see clear over the top of his head. But what he lacked in height, he more than made up for in his voice and bearing.

That was what I found lacking in Jimmy; I can only assume that time had weathered not only his features, but also his voice and demeanour. His voice was husky and barely audible, maybe that was due to years of shouting commands across the parade ground; but he had the happy go lucky swagger of a man used to getting rather than giving orders. One thing he definitely was and that was an old soldier, there was not many tricks that Jimmy didn't know and he told me of the many occasions how he had avoided spending a night in the cells. Jimmy was also a very good Lightkeeper but being in his sixties he found it hard to do some of the physical tasks so preferred to spend his month in the kitchen. It was sad in the end that Jimmy had to retire, so with the loss of both Jimmy and John in such a short space of time; for a while the station would have to make do with Supernumeraries, later Bob Byers would become a Local

Assistant and we would have Travelling Assistants and Occasional Relieving Keepers in the form of students for cover during the absence of keepers on transfer and illness.

I spent my spare time salvaging some of the best lengths of the tongue and groove boarding that we had discarded. It would be ideal for making packing boxes for the next flit that we had to make. In all I made six boxes measuring about 3 foot by two foot by two foot deep, some I'd lined with old roller blinds; they would be useful for bedding, curtains or clothing. All of them had either rope handles or carrying handles of wood bolted onto the ends. When they were all finished I undercoated and painted them a pale yellow on the outside; it was not my ideal choice of colour but it was all I could find in the paint store that was of no further use to the Station, never one to look a gift horse in the mouth I was happy that the bold black lettering of my name stood out in contrast. Boxes like these would come into their own if ever we had to have our effects transferred by lighthouse tender, come to think on it though there were no such lighthouses left in the service so the exercise did seem a little OTT except for protecting the more valuable or fragile bits and bobs that we had yet to accumulate.

Like all Stations in the service, the lime washing or boundary wall emulsioning was done on an open works order so it could begin at anytime the weather permitted without the approval of the Superintendent. All of the other painting was done after he had visited the Station and assessed the need; the earlier he came and the more additional work was found the more likely it was that the Commissioners would make a visit later in the year. He was not at liberty to say directly if you were due a visit but all the indications were there; especially if you were not visited the previous year.

With InchKeith being so close to Head Quarters, it would not be unfeasible for them to make a surprise visit but it was thankfully unlikely, the Commissioners had other pressing engagements with their normal day to day tasks and they only assembled once a year for their annual tour of inspection. On the occasion of Her Majesty holding a Spit Head Review, the Commissioners would be obliged to board the Pharos and attend, forming part of the line of ships for Her Majesty on board The Royal Yacht to review in turn.

Whenever the Pharos sailed the grapevine began, many times we would be the instigators because we could see the movement of the Pharos the moment she entered the lock gates at Leith. If she was flying the blue ensign she was on normal duties, that didn't mean that a Superintendent or the General Manager was not on board it just meant that a commissioner was not, on the occasion when the commissioners were on board their pennant would fly from the mast or their ensign from the sternpost if she was at anchor.

The Keepers at InchKeith were exempt from dipping to the Pharos on her normal duties; it was the same for when a Royal Naval ship passed us on the way to or from Rosyth. Bob said, "The flag would up and down more times than a prostitutes knickers, and we'd have no time for other tasks." We would of course still fly the flag on Sundays and we would dip the flag to the Pharos if the commissioners were on board.

The only other time we dipped the flag was when the newly commissioned aircraft carrier HMS Illustrious sailed up the Forth for the first time.

The Pharos brought out the Superintendent and Bob and I would go down to the landing to meet him. Sometimes it would be Archie Grey and the other times it was the newly appointed Graeme Simpson, first time round I believe it was Archie; though I'd hate to be quoted on it as fact. What I do remember is that both Archie and Graeme were reluctant to travel up to the Station on the tractor and preferred the walk instead. I would leave Bob and the Superintendent to make their inspection of the Station; this was an occasion when all the keepers had to be in uniform, for Bob and I that would mean the full dress, just as we would for relief days. For Jimmy or John that would mean a boiler suit or jumper and trousers and their hat.

We always had the Station ready for the inspection, so we would not expect to receive any adverse comments on its cleanliness; nevertheless just like the Navy and the Army for that matter, it is the prerogative of the inspecting officer to find fault with at least one small item no matter how insignificant just to keep us on our toes. The work that he gave us to do in my first year was to paint the inside of the tower, paint all windows and doors on the inside of the buildings and paint the receivers and foghorn. There was an intimation that the outside of the tower block would be painted next year along with the exterior of the buildings he also said for us to expect a visit from some contractors in the summer, who would prepare some of the tower rooms for use during the automation program. All of his comments would be noted by Bob and written in the monthly returns book; Bob would also acknowledge his visit and his comments in the monthly letter, but before any contractual work was undertaken, confirmation would always be notified in a Station Letter, such intimations were also made general in a Station circular sent to all Lighthouses and Shore Stations. It was a bit like the Military, if they thought you should know, they made pretty sure you did know so that there was no excuse for ignorance. It was incumbent upon all soldiers to read Regimental Orders and in my case Battery Orders; just as the keepers would need to read Specific Station Letters and Station Circulars.

Of course with his visit and works order schedule for this year and the forthcoming program for work to be carried out by the contractors; it would make it almost certain that the Commissioners would not be visiting us this summer.

We began as normal when the first heat of spring warmed the air and this year it was about the middle of April, the kind of emulsion we used for the boundary walls was called Solidec, the ten litre tubs were concentrated emulsion so we needed to add water to get it to the consistency of batter, the problem was that by adding the water you increased the risk of the batch freezing before you could apply it to the walls; hence the reason for waiting till the sun had a bit of warmth to it.

There was not much to the emulsioning, the worse problems we had was getting the odd jap in the face and hair, stiff joints from the chill of exposure to wind and as in my case, getting a squirt of the foulest liquid I had ever experienced from an irate fulmar; who no matter how much I tried to assure her that if she didn't spray me I wouldn't drop paint on her. She got the better of me by flying off before I could retaliate. Being attacked by birds would be a regular occurrence not so much up at the station; that was far enough from their nesting sites so as not to bother them, but when you went towards the gun emplacements you would get buzzed by the herring and blackback gulls, the further south you went the more likely you would encounter the terns. They would do much more than just buzz you, given half the chance they would actually make a strike; it would take someone foolhardy not to take precautionary measures, I on the other hand would rarely venture into their territory without good cause and without a helmet.

The nesting season was also the time for us to do a good turn for the Pilot boats crew as well as help Edinburgh City Council who had deemed the increase in the population of gulls out on InchKeith a positive health hazard, not for us but for the residents of the city. The report has it that they were in constant fear of attacks by marauding gulls and that the refuse tips were covered by the marauders in search of food. The crew on the other hand found gulls eggs to be quite a delicacy so we gathered freshly laid eggs at the beginning of the season. Even the crew did not relish the thought of finding something other than yolk and albumen in their eggs; perhaps we could have sent them off to the orient where I'm reliably informed such delicacies are worth a fortune although by the time they reached their destination it would probably be fledglings that emerged and not eggs. I didn't think it made much difference to the population of gulls as they seemed to lay another clutch within days it would take much more drastic action than that and we were dreading the powers that be from introducing the likes of rats or weasels to the island. The rat would prove a greater danger to us especially if their urine was to infect our water supply.

The Superintendent had sent out some portable staging so that we could paint the inside of the tower. We had discussed the prospect of putting in a bid to paint the outside of the tower next year. Duncan Leslie had replaced Alec Dorricot and we were all up for the task; no matter who was out at the Rock at the time. We would stand a good chance of getting the contract as it had up till very recently, been the Keepers who traditionally did all the painting at the station. Given the nature of the tower and the work involved there would be little risk to our safety and more than that, it would be far cheaper to employ our services than that of a contractor. We were successful in our bid and I believe we were the last Keepers to paint the tower of a lighthouse in Scotland. There I go again using the term we, as it so happens Bob and I were ashore when Duncan and Dougie painted the tower. They had to use a Bosons chair tied and suspended over the balcony rail, raised and lowered by the other keepers as one poor bugger swung back and forth with paint bucket and brush. I was only too happy not to have been that poor bugger because I know who would have used his term we and it would be me doing all the swinging.

Duncan and Dougie had done the hardest part and we were left with all the trimmings, such as painting the dome and astragals as well as all the windows and doors of the engine and accommodation block. Painting the dome meant wearing the safety harness, I chose to wear it so that I would have two anchor points on a short tether allowing me to move from one position to another yet always have at least one anchor secured, the short tether meant that should I slip it would only be a short drop and I should be able to pull myself up. Bob on the other hand was of the old school of keepers and would not have worn the harness at all if I had not diplomatically said that I would not like to have to explain to Chrissie the events leading up to his fall.

Bob had spent much of the service at a time when health and safety did not seem to take prominence, however recent legislation meant that if the service provided you with safety equipment then you were obliged to use it. I would have been loath to attach the safety harness to him after the event of a fall and in the end Bob would not have liked to put that kind of imposition on me. So common sense prevailed, though I did not quite see the adequacy of the single anchor method he had chosen to adopt even if it was one that was advised in the instruction manual.

We had the contractors out for a couple of weeks and at the same time the artificers had come out for the annual service of the engines, so it was quite a busy place and had probably not seen so much activity since the days of the garrison. One of the major problems of having so many people on the rock was the water supply. The contractors would need a supply for their building work as well as for all their personal requirements. It was not that the supply on the island would run out it was getting access to it from the tank to the tower. The contractors landed a bowser on the island at the same time as a cement mixer solved that particular problem. It was up to Bob and me to come up with the solution of uncoupling the overflow of our water tank and attach an alkethene pipe to it. That way we could fill up the bowser without having to tow it down to the well, more than that we would always ensure that our tank was filled first. The contractors would pay us a nominal sum for doing what we would be doing as a matter of course. They would be responsible for providing food for themselves but if they ran desperately short of anything then we could give them temporary loan of our stocks, whatever we supplied them we got back three fold. They would also use our suppliers for groceries and meat and we let them phone in their orders; we would even take the tractor and trailer down to the landing to pick up the stores from the Sea Hunter. She was an old wooden trawler that had lovingly been restored to her former glory, she was not an official boat for InchKeith but I used her hire to take me and my brother and his family out to the island when he came up from England for a visit. The contractors used the services of her crew a couple of times a week, either to get provisions or as transport ashore. I'm sure the crew found this time most rewarding and were as sad as we were when they had finished the job.

The Station was looking really good now and the Superintendent gave his approval of our efforts, this year we were sure that the Commissioners would pay us a visit.
There was one sure thing unless something tragic was to happen to all three Lighthouse tenders, we would be the first on the Commissioners voyage.

The Pharos was spotted in the lock at Leith and through the telescope we could see that she flew the Commissioners pennant; we would have time enough to raise the flag and do the last minute checks then make our way down to the landing.

Everybody and that included the boats crew looked dapper in their best uniforms, there were seven in the Commissioners party that included the General Manager and the Secretary, some wore blazers with the ensign motif sewn onto their breast pocket and some wore hats similar to our own but with the ensign for a cap badge. Judging by the wear of the hats, it was obvious that they had held the position of Commissioner for some time. Once they were safely landed ashore, Bob formerly introduced me; Commander McKay remembered me so held back from the introductions so that he could take me to one side and speak to me a little less formerly.

That was a nice touch and made up for the lack of ceremony on my appointment, so he kept his word, when he said that he looked forward to seeing me once I was appointed. It was a nice sunny day and only three of the Commissioners and the Secretary climbed onto the trailer for a lift up to the station; Bob gave me strict instructions not to take any unnecessary risks and to make the trip up the hill as smooth as possible. He need not worry on that part, I had no intentions of endangering my passengers, and I even went as far as to remind the Secretary to sit the other way round when he came perilously close to trapping his legs between the back of the tractor and the trailer when I had to manoeuvre round the hairpin bends.

Bob could only tell me how things went after the Commissioners were safely aboard the Pharos and making their way to the next lighthouse on their voyage.

It seemed like a lot of fuss over nothing during their visit but rest assured they were all astute enough especially the General Manager to know if something was not quite right at the Station or if improvements could be made here or there. It seemed more that just coincidence that within weeks of their visit we were to get a new tractor when the old Massey was in perfect working order; but then again the old Massey did not have a safety cab or roll bars and perhaps it would fall foul of the new health and safety regulations. Whatever; a barge was hired to transport the new yellow International tractor to the island and the Royal We was to drive the tractor off the barge. It wasn't me who had to sign for the tractor so on Bob's head be it; normally it would be the responsibility of the deliverer to land the tractor, but for some reason no-one could be found who could drive it. I can only assume that the tractor was put aboard the barge by crane or perhaps it was because nobody wanted the responsibility of driving the short but hazardous few feet from barge to the landing. I was none too keen on the idea myself but we couldn't stand around all day pontificating as the tide was already on the turn and soon the deck and the landing would be at different heights making the transfer even more hazardous. The danger was that the tractor might push the barge away from the landing when the front wheels were on terra firma and the rear wheels still on the barge. The crew with Bob's help held the barge as tightly to the landing as possible, I started the tractor and began the manoeuvre; I made the decision to complete the manoeuvre as quickly as I could in the hope that it would all be over before the barge had time to move. I could see Bob take a sharp intake of breath when I began; and exhale in a sigh of relief when the job was done. Bob knew I was lying when I said

"Piece of Cake," but he also knew that it was not directed at him but the cowardly crew.

I had a brand new toy to play with but I was still a little sad to see the old Massey go; I took the same care in parking her safely aboard the barge as I had done with getting the International off. To be quite honest, apart from the cab there was not a great deal of difference between them, in fact nearly all of the additional features were quite useless on the island, so the novelty of the new tractor soon wore off.

Chapter XII
The wreck of the Swither and Helicopter Relief's

I was quite an old hand now and on the morning of the relief's I would turn the radio on and look out the back garden; through the trees I could see the Forth and the outlines of Oxcars lighthouse; I was able to hear and see what kind of relief it was going to be. Before the shipping forecast came on, I heard the Scottish news and was intrigued to hear that the Fishery Protection Vessel; The Swither had run aground off the southern tip of InchKeith. The remnants of the storm that brewed up yesterday could still be felt and looking out at the Forth I could see that it was going to be a choppy crossing; but I had seen worse, so I could not have imagined what kind of catastrophe had befallen the Swither, forcing her to run aground. One thing was for sure I would be able to gauge for myself in the next few hours. I was hoping that Margaret hadn't heard the report; it would only have made her more anxious than normal on relief days. The cat came out of bag when Margaret joined me at the back door and Bob just happened to come out too. "Did you hear the news?" It was too late now; but Margaret did not seemed too perturbed as both Bob and I made light of the subject and we passively remarked that it might be a little choppy going out.

Choppy was the understatement of the year; I estimated that there was at least a six-foot swell as we climbed down the Jacob's ladder into the launch, timing and balance were going to be critical to avoid ending up in the ice cold water. Bob and I both looked a little green about the gills and were speechless until our feet were on solid ground. Thank God that landings made by Beecher's Buoy were a thing of the past and completely unnecessary at InchKeith otherwise I might have chosen another occupation. As it was we were both in need of a cup of hot sweet tea so we did not wish to teagle to long at hand over. Duncan relayed to Bob the known facts about the Swither from the Lighthouse point of view so that the necessary Wreck Return could be filled in. Bob could also update the Return as more information came to light. It was probably the worst day of all for an incident like this to happen as far as we were concerned because Duncan and Dougie would have enough worries getting ready for the relief without having to gather the information needed to fill in the Wreck Return.

Once we had unpacked and had our hot beverage, Bob and I made our way down to the south point. From the Station you could make out the Swither as she was held fast on the jagged rocks. We had seen the Swither many times as she cruised up the Forth and it was not uncommon for her to take the narrower south channel to gain entry to Leith. It was sad to see her in such a forlorn state as she was trapped from her bow to almost amidships in the firm grasp of the rocks. She was a steel hull trawler built just after the war and her design was similar to the armed trawlers that acted as convoy escorts and patrol vessels during the Second World War. She had served the nation well during the Cod War with Iceland; but she was coming to the end of her working life to be replaced by a more modern vessel built specifically for the task.

The longer she remained on the rocks the more weathered her sleek gray lines would become.

The news reports said that the intention was to salvage her by re-floating her at high tide. Both Bob's and my opinion was that she was held far too tightly for that and even with the assistance of a tug pulling at her stern she looked destined to remain a permanent feature on the Island of InchKeith.

The Forth Ports Authority had other ideas because they were afraid that she might slip of the rocks during a storm and sink in the channel, blocking it to traffic; once the decision had been made that she was no longer salvageable, she would be stripped of all valuables and depth charges would be set to break her back. It was hoped that she would join the other wreck that lay less than fifty feet from where she sat at the moment and so pose no problem to traffic using the south channel.

The Ports Authority asked us if we could oblige by keeping an eye on her; it was expected that she might draw some attention either out of curiosity or for a criminal purpose. It was Supernumerary Robert Boyd who noticed something strange onboard the Swither; it was around lighting up time and he was up in the Lanternroom. There was smoke coming from the deck at the stern of the vessel and he could just make out a couple of figures on board, similarly the shape of a smaller craft was seen as she bobbed up and down behind one of the other rocks. His immediate thoughts were that it was youths up to mischief; but reckoned that Bob ought to be informed and take whatever action he deemed necessary. Initially Bob and I concurred with Roberts's summation; but on closer inspection through the telescope we could see that the two guys on the Swither were taking it in turn to wave their hands in an effort to attract someone's attention. Not the actions expected from youths out for a lark; nevertheless, whoever they were it was obvious they had no legitimate business onboard the Swither. Their now obvious calls of distress demanded that Bob take immediate action in phoning the Port Authority; whose own craft was similar to the pilot boats and it was not long before they were pulling alongside the disabled boat and taking off its sole occupant, the coxswain at the helm of the motor launch did a magnificent job of manoeuvring his craft close enough to the Swither without meeting the same fate. We could only imagine what was taking place on board but it would not take much thought to picture a couple of burly police officers with hand cuffs at the ready on board the launch.

The police were kind enough to let us know the outcome of their enquiries and to inform us that our evidence might be called upon in a court action. They thought it unlikely though because the guys had confessed and admitted they were up to no good. They had hoped to find something of value still on board but found themselves marooned when their boat's engine packed up. It would be less than an hour before darkness fell and their families would naturally become worried at their non appearance, so they would be forced to go to the police; letting the cat out of the bag in the process. No matter; what they needed most was to be rescued and have to face whatever would befall them as miscreants. I can imagine their relief at being rescued and upon seeing the boys in blue; would be uttering the immortal phrase, "It's a fair cop guv."

That would not be the end of the Saga of the Swither; her presence on the rocks still posed a danger. They still proposed to explode depth charges in her bowels to break her back; acknowledging the fact that we might be in danger from flying shrapnel they thought it appropriate to notify us of the exact time of detonation. We thought it best to seek shelter in the gun emplacements being as they were built for just such a purpose. Bob had a call from the depot; the Marine Superintendent wanted us to inform him after the explosion, the status of the wreck marker buoy because it was so close to the Swither. The Port Authority gave us the time of 9 am as the time of detonation; that was good for us because we would not have to alter our breakfast schedule and we would have plenty of time to get to the emplacements.

We were still at breakfast when at twenty-eight minutes past eight by our clocks there was this muffled whoomph. We all looked at each other in disbelief, then relief when there was no cascade of flying debris or shock wave following what could only have been the detonation. "Was that it?" I said to Bob, "Must be" he replied, it was confirmed a few moments later when the telephone rang; it was the Port Authority enquiring as to our well being. All Bob could say in reply was "What happened to the 9 o'clock schedule?" The phone went quiet for a moment and somebody more senior came back and apologised profusely for the error. Bob felt like saying "Tell that to the bereaved" but thought better of it. In the end we were quite disappointed, if we had known that there would have been such an anti climax in the estimated power of the explosion, we could have watched the spectacle.

The telephone rang again shortly after only this time it was the Marine Superintendent enquiring about the buoy. We hadn't had chance to examine the damage yet let alone see if the buoy was intact. Not once did the Superintendent enquire how we were; so I say to Bob jokingly, "Tell him its fine only it's now sitting just off of the landing." Bob took a while to compose his words and in the end he settled to telling him that he would phone back in about a half hour once he could assess its condition.

Robert was on watch so he would have to wait till later to see the outcome. Even from the Station you could see that the Swither was still firmly attached to the rocks only now she her stern was at a different angle to the bow. Any other damage seemed superficial but the object of the exercise had been achieved; success would come at the next high tide when her stern would slip gracefully back under the sea.

I doubt that anyone was witness to this because it happened during the darkest hours; we noticed it the next day and made note of it in the wreck return. The last entry on the Swither was made when the rocks grasp was so weakened that she finally succumbed to the power of a storm and slid off the rock to rejoin her stern at the bottom of the Forth

Being relieved by the Pharos was not a problem for us, indeed it was not a problem for the Bass Rock or the Isle of May either; however, the Bell Rock was a different kettle of fish altogether. Relief's there were dependent on the tide and the state of the landings. They were becoming increasingly delayed because the Pharos could not make a landing; consequently the helicopter was called in to make the relief the next day. InchKeith would sometimes act as a transit camp for the Keepers for the Bell Rock and the helicopter would land here to pick them up. Maybe these were early symptoms of climate change and an early indication of Global Warming but in the eighties people seemed less interested than they are now. A decision was made that the Lighthouses of the Forth would all be relieved by helicopter; the trouble was that in order to accomplish that, a consensus among keepers would have to be obtained. Some of the keepers would lose out on a week's shore time. The relief's on the West Coast coincided with our own so it was impossible to schedule the helicopter in for that week, a much more expensive alternative would be to charter a second helicopter. The Keepers who chose to go out a week early would be compensated monetarily as their lost time ashore could not be repaid, any Keeper choosing to stay ashore for their full month would not be penalised and his place would be filled by a Supernumerary until the normal relief day; the exception being the Keepers of the Bell who were unanimous and all preferred to accept the change rather than the more iniquitous status quo; the major beneficiaries of the change would be the Keepers out on the other Rocks who would be relieved early and so serve a week less on the Rocks.

The use of the air-band frequency was restricted to trained operators; it was not essential that we as Keepers needed to contact the pilots directly but it would be an aide to them and so make for a much safer relief. To comply with that aspect Bob and I had to go to the depot and attend a radio course; on successful completion we would each receive an inshore operator's license. That permitted us to use the air-band and under our license train others to operate the radio as well. We had to do this during our time ashore so the Board duly compensated us for our time and paid for our licenses that had an expiry date that exceeded our retirement from the service.

Instead of going to Leith the limousine would go in the opposite direction to the airport at Turnhouse; because of the weight and capacity restrictions, we were also sent red plastic boxes that were used for helicopter reliefs. We had two boxes each and there were six boxes for provisions, the grocer and the butcher might not need to use all of them but if they did then that would be about as much as the hold the Bolkow could carry, leaving not much room for anything else.

We were issued hi-vis orange survival suits; they were sealed at the neck and wrists with rubber and zipped up the front with a formidable zipper that both sealed and zipped at same time. On the upper part of each arm and just above the ankle were vents that allowed the suit to breathe but become sealed on immersion. It had a tightly fitting hood; at the other end the feet gave it the characteristics of a romper suit. I tried it on for the first time and frightened the life out of the kids, Margaret on the other hand looked on in mild amusement remarking that it suited a big kid like me, but she didn't like the colour.

With my uniform on underneath the one-piece survival suit; I could rest assured that my chances of surviving an icy dip were greatly enhanced. Well that was what was intimated in the instruction literature, I was more concerned that I was going to melt before we ever got to InchKeith; that was until after the customary safety video, we learned that we would be first. This would be the first time that Bob experienced a helicopter relief, and to complete a day of firsts this would be the first time that I would take the lead.

I could tell that Bob had the uneasiness of somebody not used to flying; and he never spoke a word until we landed at InchKeith some twelve minutes later; even when the pilot asked the direct question "Are you both OK". His voice came through the headsets that both Bob and I were given; I acknowledged the pilot but Bob remained inanimate. In case the headset was not working the pilot turned round and gave us the thumbs up, this time Bob responded with his thumbs up in return.

Duncan and Dougie had prepared for the changeover to helicopter relief by painting a reflective yellow circle some eleven feet in diameter with an H in its centre the legs of the H pointed toward the north. Instead of the lighthouse flag flying from the flagpole, there was a windsock. It was even more important now, that the status of the Station was reported the night before to the relieving Principal, because there simply was not the time to do it on the relief. There was a serious debate as to where the helicopter pad was to be sighted and test flight was organised with Bond's senior pilot to select the best landing. When it came down to it there were only two choices; the courtyard offered the best site but there was the danger of hitting the suspended aerial that went from the tower to an anchoring mast, it's traverse took it diagonally across the back of the courtyard and would only really pose a problem to the pilots if there were heavy cross winds on landing; in that event they could make an alternative landing on the next flattest piece of ground toward the foghorn. The problem with this site was that even with a purpose built pad, the surrounding area might kick up dust that would be thrown into the rotor blades.

There was only one time that I felt that we were ever in any danger and that was when I was going out to InchKeith as Keeper in Charge.

Bob only made two reliefs after the changeover before he became ill on the Rock and had to go ashore, he was destined not to return to InchKeith. I was sad to see Bob go but was happy that he would see his last years before retirement at Tiumpan Head on Lewis, not far from Stornaway; but almost as far west that you could get from his hometown of Boddam on the east coast just north of Frazerburgh. Nevertheless it would be a land Station and he would not have to make another relief. Bob had been a good colleague as well as being my Principal. He had taught me well and I was well prepared to take the reins so to speak until his replacement came along.

The most endearing thing about Bob was that he always allowed me my scope and when I came up with good suggestions he was eager to accept them, on the other hand though, I would be admonished if he thought them bad with the immortal phrase, "Is the manager in?" Bob was equally nice ashore and Margaret and I could not have wished for better neighbours. Margaret was only upset with Bob once; it was just after New Year and we had just come home to the Crescent, normally I would go straight to bed for a couple of hours before the girls came home. This was like a wind down time and helped me to cope with the bustle that the girls would throw up. This time though I was just about to settle down when there was a knock at the door. I simply could not refuse Bob's kind offer of seeing the New Year in traditionally with a glass of whisky. Problem was that Bob kept filling my glass, I had no idea how much I'd had to drink, but it did not take long for the effects to kick in. I've never been much of a drinker; well not since my army days, and Whisky was never my favourite tipple. I retained enough of a sobriety to know that I would soon be throwing up over Chrissie's lovely living room carpet if I didn't act soon. Maybe it was my pallor that gave Bob the idea I was about to do just that. I can't remember staggering to the door but I do remember Bob's supporting arms guiding me beyond his threshold. There were no comments later to indicate that there was anything out of the ordinary about my behaviour so I must have made my way to our front door without drawing any attention.

Luckily for me the front door was open so I made a bee line for the toilet, I'm sure that my cries for Hughie and Ralf did not just attract the attentions of Margaret and that half the Crescent must have heard. It was two days before I fully recovered and was in a fit state for Margaret to welcome me home from the Rock. She paid me back in the end by doing a Pickford in the bedroom; changing the position of the bed just before I was due home without telling me. Well it was my own fault really if I had drawn the curtains that morning before trying to get into bed I would have known the bed wasn't in the usual place and so save myself the sore bum and the humiliation of landing unceremoniously on the floor.

I was now going out as Keeper in Charge of InchKeith and I had the company of Paul Mann, yet another Supernumerary and the youngest Keeper in the service but he came from good lightkeeping stock. There was not much I needed to teach him and I felt as confident in his abilities as I did my own.

There had been thick fog in the Forth and the remnants of it had settled round InchKeith, nowhere else just InchKeith; nothing of the island could be seen and the whole scene took on an eeriness that reminded me of the film the Trollenburg Terror; imagining that we would land to find the partial remains of Duncan, Dougie and Bobby Byres who had at long last been appointed as a Local Assistant.

Without really thinking about it I remarked "Now somewhere under that is InchKeith." The comment was intended for Paul but it was the Pilot who answered rather tersely "I know perfectly well where InchKeith is." Feeling well and truly rebuked I sat back and kept any other thoughts to myself.

I could sense that the pilot was having difficulty finding his bearings, the fog was still thick enough to obscure any landmarks anyone unaccustomed to the island would recognise. He had over shot the Station and was making a gradual decent on the east side more or less at the same height as the gun emplacements and just below the Station.

It was at this point in a more respectable tone that he asked "Where are we?" Those words restored my respect for the pilot; I'm sure that many a pilot would have turned back at that point; but with those words I knew that he was intent on doing his best to complete the relief. To best help the pilot I added the suggestion that we keep at the same height and make our way to the north of the island toward the fog horn; it would be easier to find the alternate landing from there if he chose to and secondly the fog horn was just audible over the noise of the rotor.

The pilot felt a little more confident now and thought that he might try for a landing at the Station, I thought him a little over confident at this stage because the Station was still obscured by fog and only the assistance of a breeze generated by the rotors was there any indication that it might be lifting, wisps of fog broke suddenly to allow a brief glimpse of what lie beyond but not far enough for any real definition. Once again the pilot had overshot the pad but at a height where he was not in danger of crashing into the Station and outbuildings or the aerial. He opted to try a hazardous landing on the tufted uneven ground just beyond the Station. Out of experience I knew that the ground in places seemed deceptively flat but the deception in this instance might prove fatal, and I expressed my concern to the pilot. He used his skill to hover eight feet above the ground before coming to the same conclusion. It really looked as though we might have to return to Turnhouse and abandon this method of relief. But the pilot was prepared to make one more attempt only this time he would take the same route but more slowly and allow me to use my knowledge of the terrain to guide him to the Station's helipad. In the end it became unnecessary, we were making our banking turn to the left and the gradual climb up to the Station when a gap miraculously appeared in the fog, giving the pilot full view of the Station. Taking advantage of the situation the pilot sped up the process before the fog had chance to thicken once more; and he landed us safely on the helipad. You could sense the feeling of relief from those on the outside of the aircraft as well as those on the inside; none more so than the pilot as our safety was entirely in his hands. He turned round to me clasping his hands and shaking them in gesture of thanks and mouthing the word in affirmation. I understood the gesture and reciprocated only I added the gestures for me and you at either end, after all it was his skill as a pilot that landed us safely.

There was not time for Duncan to ask for details but I think he appreciated what had taken place and he used both hands to clasp mine in his gesture of farewell.

Chapter XIII
The Riddle of the Bones & The Winter of Discontent

The Board must have had great confidence in me because nobody had been appointed to replace Bob, the bush telegraph said that Davy Leslie was coming over from the Isle of May but as yet neither he nor the Board had confirmed it. Once more I would be going out as Keeper in charge and the night before the relief Duncan telephoned me to pass on all the details of the status of the Station.

The relief passed without incident but that would just be the calm before the storm so to speak. Just as soon as I had settled in and we were having our usual morning coffee the tale of bones began to unfold. There is a well-worn pathway that leads from the Station down the east side to the foghorn and Bob on his travels had noticed a couple of bones broaching the surface where the erosion was worst. His curiosity was aroused so he began carefully to remove the soil around them and in the process uncover some more, eventually he thought that if he continued there was a very good chance that he might destroy any evidence of foul play. He was already convinced that these remains were human. He returned to the Station and informed Duncan, naturally Duncan was duty bound to investigate further. Rightly or wrongly Duncan concluded that the bones were very old and not worthy of further attention. It was obvious that Bob was not happy with Duncan's decision to take no further action; otherwise he would not be relating it to me now. That left me in a very difficult position if I chose to do the same would Bob be content or would he go outside the service to resolve the situation.

First things first, I would have to establish for myself whether these bones were indeed of human origin. I am no Sherlock Holmes and it was not within my Ken to solve the mystery but my observations could prove invaluable in any further investigation by the authorities. I followed Bob down to his discovery.

My first observation was that apart from it not being a complete skeleton the bones appeared to be haphazard in their arrangement. That was my first clue that these bones had lain there for sometime; if Duncan had come to the same conclusion then it was possibly that evidence alone that made him decide to take no further action.

Without a skull and very few other bones to go on, there was little evidence to confirm that it was a human skeleton. That was until I saw and end of the largest bone in the collection. I would have to scrape away more earth to confirm my suspicions that this was the end of a femur. Unearthing it I had the evidence I needed to confirm that the skeleton was indeed human; not only that but judging by its length he must have been a considerably tall man. I am five foot eight and holding the bone against my own thigh I could estimate that he must have been over six foot tall. The odds of a woman being that tall are even longer than the femur so I thought I was safe in my assumption and could now give him the title he. This finding helped remind me of the story that Jimmy Budge told me and would be instrumental in my future actions. I gave him the name Ivan; I had more or less convinced myself that these bones might be the very bones that Jimmy had uncovered a few years ago or those of a very close relative. Bob was still not convinced when I told him the story and he only became content when I told him that; as a precedent had already been established I would follow the same procedure.

First of all I had to phone Duncan and let him know of my intentions; the Sword of Damocles was held above me no matter what path I chose, in the end I chose to take the path that was less likely to bring the service into disrepute and ridicule. I said that Bob was unhappy that Duncan had chose to do nothing; I would not criticise Duncan's decision openly but my thoughts were, would Duncan have taken the same action had he himself found the bones anywhere other than on InchKeith. I assured him of my conviction that the bones were human and then went on to tell him about the precedent and my intentions to follow the same procedure. He still did not agree with my decision; but I consoled him that I would leave him out of the equation especially as he failed to properly inform me of the status of the Station or in his monthly letter to the Board note the disquiet of one of the Keepers.

I telephoned the Secretary to apprise him of the details and the precedent that was already set. I strongly suggested that if he chose to inform the police of the findings that he refers them to me so that I could relay the information in detail.

I was expecting a call from the local constabulary; which in our case was Aberdour in the Kingdom of Fife; instead what we got was a police launch with a full accompaniment of detectives and a body bag. I deliberately had Bob with me so that he could now be satisfied and the case could be closed as far as the Station was concerned. Bob would have to go with them, showing them the find and go through the exact moment of his discovery then relate to the police any changes in the scene since the discovery. Before that I introduced myself and gave my overall appraisal, firstly I told them of my concerns that they hadn't telephoned me before setting off, they must have thought that I was annoyed that they had not sought permission to land here. I tried to diffuse any misunderstanding by saying that my annoyance was with the Secretary for not taking up my concerns, I desperately wanted to avoid the Chinese Whispers Syndrome but had failed badly. As a consequence I was rebuked for not contacting them directly the moment of the discovery and secondly they had arrived with the expectation of finding a body, hence the body bag. I thought that by hanging around I might incur further wrath or ridicule so I chose to remain at the landing and start the water pump.

I did have one sympathetic ear though; and that was in the form of our local bobby, who had held back long enough to introduce himself and kind of infer that they were not from his station but were the big shot detectives from Kirkcaldy.

Half an hour later they returned rather disgruntled, the bag that they carried could easily have been manageable for one officer but nevertheless it was carried with all the reverence that it deserved. The senior officer walked past without a word not even a goodbye let alone a thanks, it was left to the local to tell me what was likely to happen next. Their observations and conclusions were similar to mine but they would have to go through all the forensic stuff to determine age, sex and possible cause of death. Once those investigations were complete they would let me know the outcome and would make sure that the remains received a proper funeral. I think that maybe Bob was concerned that this man may have died without ceremony and it was that that troubled him more than the likelihood of any foul play. Before leaving the local, kind of brought consolation for the earlier rebuke when he said that one of the detectives questioned the discovery as being human at all and he thought that it might have been a seal or other marine mammal because it was said with a hint of sarcasm and more than a trace of a smile; I could well imagine the officer being required to attend anatomy classes before he went on another possible homicide investigation.

True to his word the officer telephone me a couple of weeks later, yes it was the remains of a man and yes he was tall; the estimate was six-foot five. The cause of death was not determined but it had happened as long as a hundred and fifty years ago. All the evidence tallied that this may well have been the bone's of the earlier discovery because it was not on record how they were disposed. The police had no further interest in the case because even if there were the possibility of foul play the culprit would be long dead. It was left to the local bobby to dispose of the remains.

The identity of man would remain a mystery and because name or association could be placed on him, his remains were cremated with only the police officer and chaplain in attendance. Duncan did not grudge my actions nor did he think any less of me; moreover I respect his restraint in not saying, "I told you so." That ended the saga of the bones if not the mystery.

I had been at InchKeith for nearly two and a half years and my feet were beginning to itch for pastures new, Margaret felt the same. It was not that we were unhappy in the Crescent or that I was uncomfortable at InchKeith it was more a kind of sixth-sense that said that our time in Edinburgh was drawing to a close. To circumvent that and to get the ball rolling I requested to be considered for transfer at the end of my third year. Perhaps it was already in the mind of the Board and somehow both Margaret and I had been able to detect it because within weeks of my request we I was informed that I would be going out to InchKeith for the last time and I should prepare for transfer to the Mull of Galloway when I got ashore. Confirmation of the date of transfer would be made in writing later and at the moment this was just an intimation of intent. Either way, we were excited, Margaret more than I because for her it would be the fulfilment of a lifelong dream.

I still had one more month as Keeper in Charge; Davy had been confirmed as the new Principal but had not taken his position because of a situation out at the Isle of May that required his experience to resolve. It looked very much like I would not get to serve under Davy, no disrespect to him; because I believed him to be a fine Principal nevertheless I was not too upset by the prospect.

The hope for a quiet last month looked promising but the heavens had other ideas, it would be Robert Boyd's second time out with me and to be honest I could not have had anyone better to share the load so to speak. It was not unusual to experience a few light flurries of snow especially in late January early February, but to endure four days of intermittent heavy snow followed by a freeze with temperatures falling below -20 degrees was almost unheard of and I think that period of inclement weather still holds the record for the lowest temperature. Whatever, that week would be the busiest in my career and tested not only my lightkeeping skills but also my abilities as Keeper in Charge. At first the snow posed no major problems; it would be another ten days before the helicopter was due back to make the relief and there was no real necessity to take the tractor down to the landing.

The problems came later with the big freeze; first of all the aerial cable was in jeopardy of collapsing under the weight of the four inches of frost that had collected over its entire length, they say that for every one inch of frost there would be a temperature drop of five degrees as there was the four inches we came up with a temperature of minus 20c, the alcohol thermometer in the Lanternroom had ceased to register beyond minus ten so could not be relied upon. Never before and for that matter never again would I see the panes of glass in the Lanternroom both inside and out thick with frost, even during the daytime when you would have thought the sun's rays would have fought through the cold and eventually win the battle. What made the situation in the Lanternroom worse was that the light was paraffin and therefore the heat being generated was creating moisture that turned to condensation worsening the condition.

In normal circumstances you could open the vents in the Lightroom in the hope that added ventilation would lessen the condensation, but these were extreme conditions.
I even put extra heating up in the Lanternroom to melt the frost to no avail. It was far too dangerous to go out onto the frozen platform to scrape the frost from the panes of glass, and furthermore the exercise would be futile because as soon as you had finished the frost would reform and you would have to start all over again. For the moment there was nothing we could do about the Lanternroom but there was something we could do about the aerial cable and it needed doing now.

The only way to clear the frost from the cable was to lower the mast that secured it at one end. This would have been a difficult task to undertake in the summer but nigh on impossible without damaging it or the cable in the process in these arctic conditions. I know that I sometimes have the habit of giving way to exaggeration but on this occasion it might be an underestimation on my part. The mast was jointed a third of the way up with its upper two thirds securely clamped and bolted in a cleft cut into wider bottom third. The plates and bolts securing the two sections were at a height extending from about four feet to about ten feet; the total height of the mast was about twenty-three feet. With our hands nearly at the point of frostbite you can forgive me for not taking the time to measure it. We could not simply undo the bolts and let the mast fall gently because it must have weighed a couple of hundredweight, more than the three of us could safely handle; no what we needed to do was secure a rope to the upper section and find a good anchoring point to pass the rope through. We did have a rope store so finding a rope long enough wouldn't be a problem, it took longer to find the block and tackle that I was hoping to find and when I did find it, it was not in the best condition; beggars could not be choosers and time was running out. The cable looked as though it would give way any moment and we still had to find a suitable anchor for the block and tackle before we could even think about undoing the bolts.

We could not just use anything; not only did it have to be strong enough for the task but it also had to be at the same angle as the drop. I was on the verge of giving up because unless we could find a suitable anchor the game would be a bogey and be far too dangerous to contemplate. In the thick grass we located the base plate of a similar mast that had long since seen use, it may have been a little rusty but it was strong enough for the purpose and we secured the block and tackle to a shackle that we had bolted on to steel mast supports on the base plate.

I think I have been using the term We in its Royal context because most of the work especially the part where I had to climb a ladder to secure the rope to the pole had to be solely my responsibility, if there was any damage to either of us or the equipment then I would carry the can. We had to take a break every so often to thaw out, it was during one of these breaks that I hoped the cable would indeed fall down and save the trouble but alas we returned to find it still supported the weight of frost even if it sagged just a little lower. We redoubled our efforts until the final nut came loose, I asked Bob to help Robert take the strain on the rope while I removed all except one of the thick securing bolts; this would act as the pivot and the point of no return, either the boys would be overcome by the weight and the mast would come crashing to the ground or they would be able to lower it gently. There was no way that I could lend them support in time if the strain took the pole beyond the perpendicular; to all our relief she held until all three of us could take the strain. With a reluctant shiver it relaxed it hold and we lowered it gently onto the support of the trailer that almost matched to the inch the height of the pivot bolt, keeping the pole horizontal.

The minute that gravity took hold on the cable as it went from horizontal to near vertical the frost and icicles began to slide off; what remained could easily be brushed off. Getting the pole back up again took less than half the time but after we had finished we were all tired to the point of exhaustion and almost frozen to the bone.

While Robert and Bob were either having a bath or heating themselves by the roaring fire, I had to see if our efforts had been rewarded; apart from the imminent loss of the cable what had prompted me to take action was the fact that the frost had somehow prevented the radio beacon's signal to be transmitted. I breathed a sigh of relief when the familiar signature in Morse code was heard on the radio. All I had to report in the station letter was the initial loss of signal, the probable cause and our efforts to restore it. The same could not be said for the light though; after the third day of the extreme weather I had to report that I felt that the efficiency of the light was somewhat diminished on the grounds that the frosty and frozen panes would make the light appear diffused, furthermore all attempts to rectify it were futile without additional heaters that the station didn't have. What I was actually doing was covering my arse if you'll pardon the euphemism, no such report had been received by the Board from passing shipping, to all intents and purposes the light must have been seen to be normal but that might not remain the case. By notifying the Superintendent I was pre-empting that notification and make him aware of my concerns; others less kind might call it passing the buck.

He had to come up with one or two suggestions that we had already tried and in the end he was loath to make special trip out to the Rock to investigate for himself nor was he going to send out more heaters because he was sure that there would be respite from the weather within the next couple of days, he noted my concerns and said to monitor the light as best we could and if I believed it to be getting worse he would look at the situation again. Just as he had predicted the weather did change as a deep depression coming from the West giving us the usual wind and rain; overnight the frost had disappeared and the power of the light had been restored once more. I still had to put my concerns in the monthly letter but nothing needed to be done about it; it was perhaps only a one off spell of weather that might be repeated; the light was to be automated within the next couple of years so after that there would be no-one on hand to do much about it let alone express their concerns. My immediate concerns were for the welfare of the crew and in traditional navy style I thought the time was right for "up spirits." I had to be careful and mind that there were no outward signs of hypothermia, we had all stopped shivering and colour had returned to our cheeks. It's miraculous what the effects a mug of hot sweet tea and the glow of an open fire can do so the drop of medicinal brandy was not to raise the body temperature as was once thought; it has since been discovered that it would more than likely have the opposite effect; no my purpose was strictly to up our spirits by downing spirits and I would record the event in the monthly letter and the Medical Return.

After every relief the Principal on the Rock would check the medicine cabinet, it was more than just a first aid kit and there were medicines in the cabinet that would only be prescribed by a doctor under normal circumstances; however, they would be life saving for certain conditions and could be given under guidance from a doctor. There were of course medicines in the cabinet that were both stimulatory and addictive; the brandy was the most recognisable. The Principal had to note the contents and what they had been used for. In the case of the brandy it always came in a half bottle and the brand was left to the discretion of the Superintendent. We would usually record it in the return as Seal broken or Unbroken, in the case of a broken seal, the reason for its use had to be justified and recorded. It would normally be replaced at the next relief and it was up to the discretion of the Principal Keeper whether the remaining contents were consumed on the spot or he retained the bottle in his personal keeping, but seldom was a part bottle returned to the cabinet. This would be my first experience of feeling the need to issue brandy, so I have Bob Duthie to thank for the level of tuition he gave me in preparation for being Keeper in Charge as no mention is given specifically to the administration of medicinal brandy. Bob could not think of a case during his time, of a keeper using either the medicines or the brandy for his own gratification; but he thought that the general rules laid down in the regulations and the common sense applied by the Principal would deter any abuse. He said it would be easy enough for anyone to dilute the brandy with water or even replace it with tea; but it would be almost impossible to repair a broken seal. We drank our measure from a bottle with an unbroken seal and we felt the glow as the amber nectar went down to warm our very core. As a tipple there were among us those who would have much preferred a tot of Whisky but I'm sure that the brandy did its job well. At least my spirit was lifted and I could face the rest of the winter a little more content. "Oh by the way, in case you were wondering what happened to the rest of the contents of the bottle", I had to keep the bottle until I was sure that the inclement weather was over and we, not the Royal We, finished the bottle a couple of days before Bob Byers went ashore.

I was more than grateful that the rest of my last month passed without incident and I shook Duncan and Dougie's hand in what I thought was a final farewell.

Goodbye InchKeith and thank you for being my second home for the past couple of years.

Chapter XIV
While you were away at InchKeith

The novelty of our new life as a lighthouse keeper's family and the new house in Edinburgh would take a while to wear off. Even the girls kept the air of excitement about it as it seemed as though every month something new and exciting would happen.

They were already making friends with the other children in the crescent but while I still had Peter at home and the new house to get into shape, I would not have time to get to know the other wives.

I loved the new house; it had a lot more character than the one we came from. I loved the bay window in the sitting room and I had a lot of plans for that room and was dying to spend some of the money we had been able to save on furnishing it; but I would have to bide my time until Jessie, Tam, Tommy and Jennifer were away and Peter was home from his first time out at the Rock. I could see the look on Peter's face when I told him of my plans; I knew he was not bothered about the money side of it, I think it was the work involved. I knew Peter well enough then to know that he didn't understand women's psyche or me in particular. I like to change things round just for the sake of it, while Peter's way of thinking was, "If it ain't broke don't fix it." It's not that he's lazy, on the contrary, I am lucky that he always takes a turn in the kitchen and does not object to doing a bit of house work; Peter thinks on the practicality of a situation and is motivated into action because of it, anything else is done grudgingly or just to please me either way I love him just the same.

We had a living room at the back of the house so the front room would become a bedroom for Jessie and Tam and once they were away, I could start stripping the wallpaper off ready for Peter to decorate when he came home; it would be just the spur he needed to motivate him whether he liked it or not.

Tommy would find a place at anybodies table; he has learning difficulties but what he lacks in one department he certainly makes up for in his social skills. Peter had been out at the Rock a couple of days and I had Jessie and Tam for a few more before they had to go back to Patna. There was a sweetie van that came into the crescent every afternoon and Tommy was already out at the van when I went to it. I could hear him chatting away to some of the other wives when suddenly I could hear the sound of laughter. I only just caught the end of the story, but apparently Tommy had gone to the van for some headache tablets and hoped the van had some, when there were none, one of the wives said she had some that she could give him; by the way she was addressing him it was like they were already firm friends. When Tommy said, "I must be going though the change of life," all of the women there laughed and had a growing liking for Tommy.

Tommy had well and truly broken the ice for me, he had struck up a friendship with Anne McIver that would last long after we had departed from the crescent and in doing so he introduced me to one of the best friends I have ever had.

Anne asked how I was settling in and all that; then she asked me in for a coffee, well I couldn't leave Jessie and Tam on their own, it just wouldn't be polite; so I returned the request and that was to be the start of it. While our men-folk were away we would spend most days in either one house or the other, our friendship was such that we would even help each other when it came to shifting furniture about or as Peter used to call it a Pickford; that was just one of the many things we had in common.

We would go together with Kirsty and Catherine and pick up Karen from the triangle club, this was a kind of nursery school organised by the community rather than the education authority. Later on Karen would join Colin at Pirniehall Primary; he was a year older than our Karen while Kirsty and Catherine were the same age and like us were best pals. We were friendly enough with the other wives in the crescent but our friendship was different. Next to our men-folk and our children we were there for each other to the point where we could sometimes read each other's thoughts and anticipate when was the right moment to go over. Peter and Donnie never shared the bond that we had, probably because they were so alike in preferring the comfort of their own fireside rather than socialise during their time at home. We respected that but it did not stop us from going out on the odd occasion when the men had other things to do.

Shopping or should I say window shopping was one of our favourite pastimes, even today Peter is reluctant to wonder around the shops just for the sake of it; but for Anne and I, going up Leith Walk, The Bridges or Lothian Road was a highlight of the week.

Once the better weather came and the schools were on holiday we would take the kids to Portobello Beach; it wasn't Blackpool by any stretch of the imagination but the sand was clean, the sea was warm and the kids had amusements more suited for their age group; with all that and the happy smiling faces of the children enjoying themselves it felt like being transported to paradise; just close your eyes and you're there. Typical of all children it wasn't long before one of them woke us up with, "Mum can I have this or Mum can we do that", soon all of them would be at it until a period of quiet reigned once more or it was time to catch the bus back to the crescent.

Leith Walk was where you could pick up a few bargains; before the days of the pound shops there were the indoor markets and these were usually owned by people from Asian community. Anne and I would make our way up one side the road going from one to the next then cross the road and go down the other side. I affronted myself one time when I mistook a man wearing a turban, sorting through some stock as a shop worker or owner and I asked him politely if he had a dress I liked in a smaller size. My face went as red as the petals on the printed dress I was holding when he replied apologetically, "I'm sorry, I'm shopping here myself."

I would cause Anne a bit of embarrassment on another one of our trips when the children weren't with us. We went into this shop and were having a good rummage; Anne found something she liked and as soon as she picked it up; a young male shop assistant swooped upon us. They were talking away when suddenly I felt the overwhelming need to break wind; I held my butt cheeks together as tightly as I could and it came out almost silently, what it lacked in sound was more than made up by the smell. I thought if I move now they would know it was I; so I stayed glued to the spot even if I found to keep a straight face and restrict my breathing to just my mouth. It was obvious that both Anne and the young man could smell it but neither let on, Anne was nudging me in the ribs and saying to the chap, "I'll have to think about it and come back later," before hurrying out of the shop and bursting into laughter. She says to me, "Did you smell that? The dirty bugger farted and it was rotten." I was expecting the smile to be wiped off her face when I told her it was I. She just laughed all the more and said I was the dirty bugger and a rotten bastard to boot because he must have thought it was her.

When we went shopping with the kids it was sometimes hard to tell Karen and Catherine apart, they were the same height and build with their hoods up; from the back it would be easy to get them confused and that was exactly what Anne did one day when we were shopping for shoes for the girls. Kirsty and Catherine had to have identical shoes after all they were best pals even if there were three shoe sizes between them; on Kirsty, Catherine's shoes were like boats. Karen on the other hand was independent at five and liked to choose her own shoes. Anne sometimes had the problem with Colin and Catherine lifting things from the shelves and it was embarrassing for Anne when she got to the checkout to find stuff that she did not want; anyway Karen was lifting this pair of shoes when she got a wallop on the back of her head from Anne. A shocked and tearful Karen turned round to reveal to Anne her mistake. Anne could not be sorrier, both to Karen and to me for taking such a liberty. It took a while for me to convince Anne that it could have happened the other way round just the same, so to forget it. With Karen it might take a little longer to forget, just about as long as it took to fill her face with an extra large ice cream that Anne treated us to.

Colin was into fancy trainers and was set on getting a pair of green white and yellow ones, Anne was not keen on them but Colin insisted so she bought them but not before telling Colin that they were the same colour as the packet of cheese and onion crisps he was eating. Colin not only liked the colours he now liked the association and that why he called them his Cheese and Onions rather than his trainers; when we moved to the Mull and Donnie was transferred to the Isle of Man, I called Anne on the telephone and it was Colin who answered; he of course was older now and liked to chat to me but there was one topic of conversation he went to; telling me that he had gone from Cheese and Onion to Barbequed Beef, the trainers he now wore were a kind of burgundy colour.

There were a couple of milestones in the girl's life that Peter missed because he was out at the Rock; one of them was Karen's first day at school. Like all first days, mine included; it was a mixture of excitement and tears when we said goodbye at the gate. That day for was less of a milestone than when Karen decided that she wanted to go the short distance to school on her own and showed the start of her independent nature. I reluctantly let her go but unintentionally followed her a few moments later as I needed some things from the shops; it was therefore a surprise for me to see her coming out of some flats. She had gone to the flats to find her pal but couldn't remember what floor she was on. I did not like the idea of her going into the flats because they had the reputation of being a haunt of drug users, so I took her to the school myself with the promise to allow her to go to school herself just so long as she didn't repeat the episode.

I was becoming more concerned about her sight; she'd had a squint since the age of two and I thought it might be getting worse, so I mentioned it to Peter. Shortly afterwards, Karen had to wear glasses and as she was the apple of her father's eye she got to pick the best glasses we could afford.

For a while now, Karen had an interest in the way she looked and was always getting told off for messing with my makeup; I would have thought she would have learned her lesson when she was about three and found some clove oil that my father had used to help with his toothache, she said "look Papa"; and dabbed a bit on her cheek, then she began to squeal as the tincture burned her delicate skin. Not a bit of it, not much later; I had to spend hours trying to wash Vaseline out of her hair; so it did not come as too much of a shock when there was a knock at the door and Jimmy Burns appeared with Karen and two other children in tow. "I think these two belong to you, but as for this wee chappie, god knows." Karen was easy enough to identify as she was the only one who was clean except for the sleeves of her cardigan, Kirsty I could recognise by her height and the fact that her muddy appearance was only superficial; on the other hand I could well forgive Jimmy for not recognising that Karen Bambrick is a girl because I doubt whether her own mother could have been able to tell. Karen was covered from head to foot in uniform mud, not thick enough to be able to scrape it off; but thick enough for us not to be able to identify her features. I knew who it was, only by the fact that she was not much taller than Kirsty and there was only one of the children in the crescent who fitted that description. Heaven knows what I could say to Margaret Bambrick; so in the end I chickened out and just rang the bell and made a dash for the house before I could be spotted. You bet your life on it that I gave my two the what for once I'd cleaned them up; I singled Karen out for special attention, in the end it did no good because she later went on to study hairdressing and makeup at college.

My curiosity got the better of me once and possibly cured me of being a nosey neighbour, Peter says that I just think I've been cured but I know I have learned my lesson; apart from the grass at the front of the house there was also some shrubs and a couple of prickly bushes thrown in for good measure. I was so intent on listening to a conversation up the street that I was not looking where I was going and ended up landing in the bushes, Anne had seen me fall and came rushing over only to find me howling with laughter one minute and greeting in pain the next; she spent the next half hour between fits of laughing and helping me clean all the prickle scratches.

Compared to Karen, Kirsty was angelic or so we thought. Karen being the elder of the two always seemed to catch the brunt of any scalding' that were going about but one day I had to rush into the kitchen when there was this almighty crash. The girls were going to make their own breakfast but couldn't reach the cereal on the top shelf of the kitchen cabinet; you know it's the type that had the sliding glass doors a couple of shelves and then a drop leaf front and two cupboards below it. The one we had was blue and as it belonged to the Board so we had to look after it. When I saw what had happened my heart missed a beat, Karen was nursing a sore head and Kirsty was nowhere to be seen. I heard her muffled crying coming from underneath the cabinet. It is amazing the strength you can muster when it comes to the safety of your children and I managed to lift the cabinet with ease, something that it would take both Peter and I to do when it came to spring-cleaning. Apart from a few bruises both of them were OK; as for the cabinet, with a little DIY, I could re-attach the hinges to the leaf and both Peter and the Board would be none the wiser just so long as the girls could keep quiet about it. They knew that Peter would be angry with them and they would end up getting a lecture on how silly and dangerous it was but they already had the incident indelibly printed in their memories and for the moment a few bruises as a reminder, so it was to remain our secret till it was safe to tell the story.

While we were in the crescent we would often go to the Church and take the girls to Sunday school; Peter and I are not devout churchgoers but we liked the atmosphere of the local church and the girls liked going to Sunday school. I would take them even when Peter was out at the Rock. Every so often the church would have a christening and the young ones would come across from the Sunday school; later on the girls could decide for themselves if they chose to be christened; but there was one Sunday when Karen had found a bottle of syrup of figs in the house; she liked the taste so much that she kept going back to it. They were just about to start the ceremony when I was called to the Sunday school because Karen was in such distress; distress was not the word for it, dat dress and everything from her waist to her heels was covered in diarrhoea. I did not know whether to cry because of the mess or cry because of how she might be feeling. I had help in cleaning her up and somebody had kindly given us some cloths, one in the congregation was a nurse and said that I should keep an eye on her and give her plenty of fluids to stop her dehydrating other than that she would be fine. I think Karen learned her lesson and both she and Kirsty chose to be christened not long afterwards.

We could not have been in a better place as far as the girl's health was concerned, first of all Karen had to have an operation to correct the squint in her eye and then the Doctor had picked up the trace of a murmur in Kirsty's heart at a preschool examination.

Karen had to have her operation in the Eye hospital at the Royal Infirmary, there were quite a few children undergoing the same operation on the same day, Peter and I took her up to the hospital and returned at visiting time then again the next day after the operation. It was a sad sight to see her with her eye all swollen and closed up and even sadder to see the other kids in the same way. The children would not be allowed home until their eyes opened and they could see out of them. In the next ward were some elderly patients who also had to undergo operations and one of them was celebrating his birthday, he was given permission to share his cake with the children and the nurses arranged a kind of party for them. What brought tears to my eyes was to see these little darlings trying to find their cake and sandwiches by feeling about the table in front of them, sadly our Karen was among them.

Kirsty was always going to be small and even at the age of thirty she's only four feet six, mind you I'm not much taller; but over the years it has had its compensations. I was for a long while able to fit in clothing made for a teenager and I'm sure Kirsty could still get away with a half fare on the buses. But back then there were concerns about her height and weight so needed to check to see how or whether she was thriving. The issue of the murmur was resolved when it turned out to be a childhood thing which would close up naturally within the next year or so and posed no danger to her general health.

The thing that troubled us was our Kirsty had already been through a lot in her few years; she was born with twelve toes, she had topic eczema for the first few years and had only just got clear of it. We used to have to put mittens on her or she would scratch herself raw, every time we changed her, her tights used to stick to her legs where the tissue had suppurated into sores, if it was bad for us to look at it was much worse for Kirsty who had to endure the pain. We thank God that it lessened through time as did the asthma that usually goes with it, neither had taken on the more life threatening forms.

Kirsty was admitted to the Western General for a week of tests; once again we had to make a trip to the hospital at visiting time, only this time the Western was just a short trip by bus. Peter was again lucky enough to be ashore this time just as he was with Karen.

We both noticed on our second visit the sign above Kirsty's bed that read "Stool Collection Daily". Peter and I both thought the same thing and showed our ignorance when we thought that the nurses had found Kirsty a wee job to do. Next door to the ward was a kind of crèche and we thought Kirsty had to collect all the chairs together. We were awfully glad our conversation had not been picked up by anyone; and our ignorance was put right when we saw another potty being collected from another bed with a similar sign. There used to be an entertainer on the television called Jimmy

McLeod and Kirsty had named one of her toys Jimmy McClown because that was what she thought he was called. She cried inconsolably when she'd left it at the hospital and the nurses could not find it. She soon became attached to another toy though to start the process all over again.

Anne and I would sometimes go over to the second hand shop in Pilton just for a rummage, but this one time Anne saw a big silver cross pram and wanted it, I thought, "What on earth does she want that for, maybe she's getting broody?", I kept my thoughts to myself just in case she would feel offended, so I limited my question to "are you sure you're needing it?" I thought it had seen better days and I had not seen one like it since I was a little girl in the sixties. God knows what Donnie must have thought when he saw it, but then again maybe he didn't see it, Anne was clever like that and she also knew how and when to twist Donnie round her little finger.

I tried some of her techniques but was only successful on a few occasions and I think that was because deep down Peter had a soft spot for animals. The girls were at the age now when they would look at a puppy or a kitten and say, "Can we have one Mummy please?" They would always wait until Peter was out at the Rock because he was adamant that there would be no pets till the girls were old enough to look after them. The trouble was I didn't want to fall out with Peter and I didn't want to upset the girls by continually saying no. It was even harder to avoid the two pet shops in Leith Walk and I think Anne was beginning to get suspicious as to why I seem to guide her towards the Bridges or Lothian Road. Grasping the Bull by the Horns I got the girls a kitten that they had fallen in love with, I would think on something to console Peter with when he came home; by that time the kitten would already be established in the family. The girls called the kitten Smokey, I could have thought of something more original but it was the girl's choice. Smokey was a common and garden type moggy, nothing really spectacular about him but he did have a lot of character and the girl's loved him. I hoped that when Peter saw him he would grow to love him too. I was still dreading the moment he came home because I thought it wise not to mention Smokey in our telephone conversations. In the end I came up with the feeblest of excuses why I bought the kitten and it only worked because I had the girls to back me up. In unison we said, "We bought it for you Daddy, to keep you company out at the Rock". Smokey did his part too by coming straight over to meet Peter. I knew we'd won when Peter said, "What's its name?" He did not want to submit straight away so he quipped, "If he's a secret five a day cat, you can just go on and take him back". The girl's were too young to understand the meaning and thought that their dad was serious; both their jaws dropped and their eyes began to well up. I stepped in and said, "It's OK girls we can keep him". Peter confirmed it by lifting Smokey and stroking him, Smokey responded by licking the fingers of the hand that was holding him. There was one thing though; Peter did not like the name Smokey, so he called him Tats, in the end the cat would answer to either but only when he chose to.

The trick we'd used with Smokey also worked when we bought a hamster we called Brownie; again it was the girl's choice of name. This time Peter accepted it because his thoughts were that the hamster would not understand whatever name you chose to call it. When I bought it for the girls I had completely forgotten about Smokey; but I could hardly take it back and upset the girls. I did not have to worry on that score because it seems as though Smokey had accepted the hamster as part of the family and treated him as a source of amusement rather than a meal. As for Brownie, the cat never bothered him and he was completely unafraid of him. I remember one time when Brownie was in his ball and Smokey flicked the ball across the sitting room floor, they played like that until the novelty wore off. When Peter came home from the Rock he accepted Brownie without objection. It was quite comical to see Brownie's antics of an evening; if he wasn't swinging on the curtains he would be sitting on Peter's lap preening himself, sharing Peter's sandwich or watching the telly with him.

The remarkable thing was it was Brownie's choice, he had found a way to get up the settee and it was his preferred resting place to sit on Peter's lap or next to him. Sadly hamsters are not long lived, so only Smokey came with us to the Mull.

Anne had at last found a use for the pram; the shows had come to Granton and we loaded all four children onto the pram and pushed them all the way. Kirsty was at one end Karen at the other and Catherine in the middle. Colin would lie in the bottom on the shopping tray. Looking back on it now we must have seemed like a family of well dressed hawkers; but at the time the kids were only too glad not to have to walk. The same could not be said for the way back as that was mostly uphill; we had to offload them in order to push the pram. The chorus of complaints of, "We're tired" as we trudged up the hill was only quelled when we came to a flat bit and they could climb aboard once more.

It was going to be sad leaving Anne and the children; but that was the way of the service, sooner or later we would all have to move to pastures new. There was always the chance that we would come together again in the future. Goodbyes are always an emotional time for me but one of the saddest moments was when I was up at the local shops and Kirsty was talking to a couple of elderly ladies; That was another thing about our Kirsty; from the moment she could talk she has been able to communicate with everyone on their level; so it was no wonder that these two ladies found Kirsty so adorable. They had often seen us passing from their window and had seen us at the church on a Sunday, now they would like us to pay them a visit and have tea with them; this was the day before we flitted to the Mull; tearfully I had to tell them," I'm so sorry, but we are moving tomorrow". Even that paled as to nothing on the day that the tragic news that Tam had passed away was brought to me. I had just returned from taking Karen to the school and I would normally go into Anne's for a coffee, she was waiting at the gate for me with a message from Peter telling me to go straight home. I knew it must be urgent and began to worry what might be wrong; all sorts of things were going through my mind. I could tell right away that someone close had died; Peter had that kind of look on his face. I loved Tam dearly and we both cried for the moment and for our loss but most of all for Jessie. It was Jessie that we had to think of now, so I went back up to the school to fetch Karen while Peter finished packing for Patna; that was

the saddest day of my life; Yes I cried when my mother died and I cried when my father died just a few years ago, but it was a different kind of grief.
Duncan would have to inspect the house before we left and to tick off the inventory of the furniture, he was due to visit us on the Sunday; and the van would arrive on the Monday. In the early hours of Sunday morning the girls had woken up and began rummaging around the house. Peter had carelessly left a Stanley knife where the girls could get it, we were lucky to find only a couple of kitchen chairs with their seats slashed and not pieces of the girls lying here and there. Peter paid for his carelessness by having to dash about looking for a store for something to cover the chairs with. If Duncan had noticed the fablon, he never said anything and everything was ticked off. Shortly after one of the nicest things to happen was; that Duncan and Catherine invited us in for lunch. But that's the kind of people they are, it was not required of them but they went out of their way to do it just the same. I left the Crescent with mixed emotions, sad to be leaving my best friend behind, sad because I would not get to know those two lovely ladies and sad because Duncan and Catherine had been so nice to us; on the other hand my dream of living at the Mull of Galloway Lighthouse was filling me with excitement; and I could not wait to see our new home.

Star of the Four Kingdoms

Like a beckoning finger come hither,
meet thy doom, ye mariners of old
My rocks await your careless gait
or stormy tempest seas unfold
No kindly welcome to Scotia's soil,
or gently rolling surf
But surging tides and jagged cliffs
topped by heather, grass and turf
They built me tall, they built me well,
my whiteness stands as sentinel
In daylight hours my shapely form
now welcomes all to Scotia

And in the darkness of the night,
my purpose truly comes to light
Benevolent flash to guide them passed
and safely on wherever
Like a Nelsons patch I 'm blinded
by my neighbour's fretful plea
that was done of purpose so I should see them not
and they see nought of me
My sweep takes me round the Bay of Luce
and Wigtowns' kinder shores
then gently on to Solway Firth
till it passes out of view

On clearest days the Lakeland Fells
and peaks my eye can see,
at night my brother lights flash dimly
if Brethren Lights they be
My sister at the Point of Ayre
has bands' of red and white;
she shows the same in flashes
in the stillness of the night
but all that distance can reveal to me
is a single flash of light
My beam sweeps on in gentle scan
to Peel that marks the end of Man

I linger just a little while,
on the fertile lands of the Emerald Isle
till blindness comes once more
as silvery sea gives way
to land on Scotia's rugged shore
Four score years I stood alone,
yet flawed by this omission
My light and form were concealed by fog
till they came to that admission
A trumpeter or Heralds host could not announce us better
Even if its sound forlorn and from your slumber
wakes and dreams all scatter

there was no better sounding Horn,
but alas once more I stand alone
Three graceful Queens
have passed me by
and Two who's Majesty gazed on me
sadly Belfast's Pride I've seen
making passage to her destiny
To all my charges great or small
and to those sons of earth who tended me,
I am simply known as ...Star...
The Light of the Mull of Galloway

©Steve Hardy, South Rhinns Community Development Trust

Chapter XV
The move to Mull of Galloway and the Station

There would be little chance of reaching the Mull much before nightfall and it would be the longest train journey the girls had taken since the one they had made coming up from England; Karen remembers little of that and Kirsty was just a couple of weeks old.

When we left the Crescent behind, we would trace the exact journey I had made on my first day as a Lightkeeper; only we would be going a good bit further by taxi to the Mull. The guys in the removal van should arrive about an hour after us or that was the plan.

The longest and newest part for them would be the second leg from Ayr to Stranraer; it was still the slow and stop for every cow journey that I remember taking three years earlier; even thirty years later it has not improved much and should another Beeching come along it would join the section from Stranraer to Dumfries and be relegated to annals of history; isolating the South West of Scotland by rail altogether. Margaret and I had other things on our mind other than looking out of the window at the passing scenery; but as soon as we got to Glenluce and the line took us out of the rolling hills and valleys of the Galloway Hills to Luce Bay now opening up before us; I could point to the merest outline pointing up from the land at the end of one side of the bay and say to the girls, "That's your new home girls"; to the reply of "where daddy, where?" Their jostling to get a better view was understandable; I knew what I was looking for, as yet the Lighthouse could have been on the other side of the moon because it looked so distant to the girls and seeing what I saw, they would think it a smudge or dead fly on the window. When Margaret looked I could almost read her thoughts before she said, "Peter, we're going to need a car." I'd been thinking that since I'd first heard that I was going to be transferred to the Mull. Asking the other keepers, I was able to glean that the Mull had not been a Schooling Station for a few years and at the moment there was only Alec and Margaret Dorricot there who had any children but both were under school age. "Hmm", I thought. The other thing I heard and could work out for myself from the map; was the distance from the Station to the nearest village and from the Station to the nearest town. "Twice hmm". Up till now I'd always tried to avoid needing to have to drive; it was the expense of running and maintaining a car that put me off, but what the heck; Margaret was so looking forward to moving to the Mull and there was no use having money in the bank if you could not enjoy it and please everybody in the process. I'd never mentioned any of my thoughts to Margaret till now, the realisation would have to come to her in order for her to accept any sacrifices that we might have to make. For now though we would enjoy the rest of the journey down to the Mull.

The 26 mile taxi ride from Stranraer to the Mull only emphasised the need for a car; to be sure there would be a weekly shopping trip but what of visiting Jessie, it was about the same distance from Patna to the Mull as it was from Patna to the Crescent but without a car it would take the best part of a day. No if my mind was not made up before, it sure was now; my first priority would be to pass my test, on second thoughts our first priority would be to arrive in one piece at the Station and hope that the van would not be long behind us.

We passed through the villages of Stoneykirk, Sandhead and the hamlet of Ardwell but it was not until we were passed Ardwell that we could see the Lighthouse once more and this time it was much closer and more defined but the girls were too tired of travelling to notice, it disappeared from view shortly after and we did not see it again until we were a couple of miles passed Drummore and on the final leg, a single track stretch road all the way to the Mull. We could see that the Lighthouse was on a promontory; on the map it looked like the end of a scrawny witches thumb only without a nail, but what we could see in front of us was the land coming in from either side to form two little bays and then the land opened out again; all the while climbing to a kind of plateau to the Lighthouse itself. The road twisted and turned as it climbed crossing a couple of cattle grids on the way. At the top, the road went across another cattle grid and through a gateway between two dry-stone walls about six foot high; these were not painted white so we had not reached the boundary of the Station even though the majestic tower could now be seen clearly; Karen was awake now and seeing it she just said "Corr;" followed shortly after by, "It's beautiful" as we turned the last bend and the Station in its entirety came into view. Margaret squeezed my hand and I could sense that she agreed, it was better than even she had imagined; even an old salt like me had his jaw dropped not just by the Station but the whole vista. There was a fork in the road but we went straight on through the gates and into the Station, the white boundary walls formed a kind of square; at the back of the square was the tower and the hemisphere of a quarter deck, behind the tower was a courtyard with the dwelling block of the Principal and the First Assistant beyond, above the doorway of each entrance there was a kind of port mantle that desperately tried to emphasise the individuality of each of the houses where it lacked other signs of detachment. A large green went at an angle from the main gates to the single pathway gate of the tower courtyard. There was a familiar building on the right; so familiar in fact it was like being transported back to InchKeith. It housed not only the engine room but also the house of the Second Assistant. Apart from not having a portico it was identical, it would all be new to the children because Margaret was the only other member of our family to go out to InchKeith.

The main reason for delay in being transferred was our house was going to be renovated but they could not start the work until the other Keeper had left the Station. Margaret and I could not wait to see what had been done, but we still had the formalities to go through. Waiting to greet us was Ken Clark and his mother, Ken waited until the taxi departed before stepping forward. He was tall and had almost white hair and he looked young to be in his late fifties but standing next to Mrs Clark you could well believe it. As Ken was tall and youthful looking Mrs Clark was not much taller than Margaret and she must have been well into her eighties; Ken was not married and Mrs Clark preferred that title rather than her Christian name; we respected her wishes and did not even go to the lengths of discovering what it might be. Whereas Alan Chamberlain had a brusque quality to his Yorkshire tones, Ken's was softer and quieter but unmistakably Yorkshire, we could not be sure about Mrs Clark but then again it was not important. Ken showed us into our house just as Bob did at the Crescent; he went through the inventory of the contents with us before leaving.

This time all of the furniture was brand new; the new modern divans came with mattresses included; the mattresses we brought from Edinburgh could go underneath adding to the comfort and making us feel a bit like the Princess and the pea, they were personal issue just like the pots, pans, duvets and bedding. The furniture in all the rooms was modern and for some reason there was a new issue of pots and pans, so now we would have two sets; It would not be quite like Christmas all over again but what made it even better for us was that along with signing for the furniture there was a Station Circular that informed all Keepers from that date all furniture issued by the Board would be the responsibility of the Keeper and henceforth would travel with him from Station to Station; all replacements needed due to damage or wear and tear should be notified to the Superintendent on his annual visit or sooner if the nature of the replacement demands it.

Now that was great news for us, we had brand new furniture and it was virtually ours.

What gave me some concerns were the Kitchen windows; nice though they were in giving a good view out to the Irish Sea; they lacked in practicality of not being made with toughened glass. They had replaced the sash windows in the rest of house but the Kitchen layout was new and so was the window frame; it had a single opener to the right of a large panoramic pane and it was as if the designer had completely overlooked the fact that the back of the house was only feet away from a 230 foot drop to the sea below and that the cliffs would accelerate the wind coming from that hemisphere. The danger was all too apparent but I would not have time to do anything about it now and Ken was not really the person able to get direct action. This was one strange situation for the service to be in. Ken could report it to the Superintendent but the Superintendent could only pass on his concerns to the Clerk of Works Projects as he would not be able to assume any responsibility until the Project was officially handed over to him, the other route for Ken to take would be for him to notify the Clerk of Works himself who would have to come out and see for himself the possible danger. I've already had one experience of Chinese Whispers and I did not Ken to have to go through the same, I reckoned I could wait until the Clerk of Works made his sign off inspection due within the next two weeks and speak to him myself.

Just as they predicted the guys arrived with our furniture and our beloved Tats who had to suffer the indignity of being sedated and put in a carrier; then found a safe place in the back of the van. I am sorry that at the time I felt it was the only way to get him to the Mull without causing him too much stress. Poor wee soul he was drowsy from the sedative and he was soiled much like Karen had done with the syrup of figs, so he was Margaret's number one priority; a task she didn't envy but one she was experienced in and Tats would be only too glad of, once he came round.

We unloaded the van in quick time; and there was still about an hour of daylight left. Ken and Mrs Clark had invited us in for dinner and judging from the delightful smell coming from her kitchen; Mrs Clark was going to treat us to something special from her long experience of the culinary art.

The cosy front room was the only place where we could eat together, ours was the largest kitchen on the Station and even that was only just about big enough to fit a table and three chairs. Later in the year they would be renovating the other houses and be repainting the tower. Ken and Mrs Clark were used having their meals on trays but she found it hard to believe us when we said that's how we preferred it too, she must have thought that we were just trying to be polite; she remarked how well behaved the girls were. The great things about the girls was that you could take them into anybody's house and know that they would treat it and the people with respect; Mrs Clark's only concern was that the girl's were fascinated with their Chihuahua called Tiny, they had lost Darragh their other one just recently and were very protective of this one as he was the same age. Sadly poor Tiny was heartbroken at the death of his pal and died within a month of our arrival; so we only got to see him once or twice.

I still had a couple of days before I was due to take my first watch; Ken needed only to give me the briefest of instruction about the light and that was all. I was well familiar with the foghorn and I could do the weather reporting just as I had done on Sule Skerry. The light was an electric array of halogen lamps that was rotated on a mercury bath turntable, just select the setting for the intensity and turn the key, there was a duplicate system on standby and they would be alternated each night. If there was a power failure or the main array failed, the last four lamps on the array were smaller and got their power from batteries. Later that year they would install a generator that would reduce the need for the standby lamps to that of catastrophic failure of the main light system. The power of the light was some 3,000,000 candelas and the light could be seen far beyond the 28 miles of its nominal range. There's a Lighthouse at St Bees near Whitehaven and it could be seen on clear nights and that was close to fifty miles away; and it did not have as much power as the Mull. At sea level the curvature of the earth would limit the distance a ship could see; however as the light was over 300 feet above that would extend the range of visibility especially at night when there were no other visible landmarks.

We would have to enrol Karen at the local primary in Drummore and see if we could get Kirsty into the playgroup. Alec Dorricot was the First Assistant and we met him and his family the next day. Because there were now two Margarets' at the Station, I will henceforth refer to my beloved by her pet name of Mags. Margaret Dorricot was going to take Paul and Stacey to the playgroup, she would introduce us to Archie Gorman and we could arrange for him to come back and pick us up. His car was not big enough for all of us and because Archie also ran the weekly shopping trip into Stranraer, he was able to give us a good discount for our other trips. There were other things besides enrolling Karen and Kirsty, we also had to register with the doctor and get some messages but thankfully all of this could be done in the village even if we had to pay a bit more for our groceries.

The village had practically everything on the one street; there were a few shops, a post office that doubled up as a registry office for the district of Kirkmaiden, a couple of pubs and the Primary School. The doctor's surgery was on the way in and just before the junction into the main street. On the way out of the village heading towards Stranraer was a small caravan site with a kind of clubhouse.

At the bottom of the main street was a tiny harbour, at the best of times I only ever saw just a couple of boats tied up to the quay but I could have imagined the harbour in its heyday. Port William on the opposite side and Drummore on this side offered the only safe anchorage in Luce Bay. Up the bay was West Freugh and out in the bay were bombing targets so the area had restrictions and it was probably those restrictions that saw the decline of Drummore.

The whole of the peninsular or Rhinns of Galloway to give it proper title was predominantly farming, any industry seemed to be concentrated in Stranraer, so the village was very rural in its outlook; there were commuters who worked in the shops and factories of Stranraer but the majority had working connections with the village or the farming community. Increasingly there were also a number of in comers from over the border who had chosen to retire or move here for its tranquillity and beauty.

I was lucky not to be included among those despite my English accent; simply because the lighthouse had been part of the community for over 150 years and in those years I don't believe there had been many Galovidians or Gallwegians among its serving keepers. I was simply known by my name or Mr Lighthouse and the same went generically although we did not live on ocean drive. The Galloway Irish accent as some of their fellow Scots described it; would be spoken only by our Local Assistant, Jock Binnie and the two Occasional Keepers, Jimmy Fyffe and Jimmy McGowan.

Whatever Sherlock Holmes there was in me was able to deduce that the Board would not have sent us to the Mull if there was not adequate provision for Schooling for the children, it turns out that recent changes in legislation meant that the local education authority were now responsible for the transport of children living more than three miles from the school. Mr Shanks ran a minibus for the school runs and we needed to contact him to pick up Karen; the trouble was with the authorities interpretation of the new legislation threw common sense right out of the window. They would not pay for the minibus to come all the way to the Lighthouse. Somebody would have to transport them down to the cattle grid at West Tarbert, one and a-half miles from the Station. That was the name given to the little bay on the left at the bottom of the hill, the bay on the right was East Tarbert; there was no habitation within a mile of that point let alone shelter from the elements and that was what would leave the authority open to ridicule. The whole point of the legislation was to limit the dangers encountered by school children going to school and from school; those having the greater distance to travel were exposed to the greatest risks and none more so than the children of the Lighthouse, who had the added danger of exposure to the elements especially the wind. Thankfully the number of times the children were exposed was limited but only because we paid Mr Shanks to come the extra distance; the authorities were adamant that Mr Shanks could not take on outside commitments at same time as the school run. So Mr Shanks sent either his wife or daughter in his own car to pick them up. The Board once they knew of the situation took over the payment until another solution presented itself.

Karen got to know the Shank's very well and even saw the horse that they kept in a stable. Without thinking too much I said to Karen that the car would not be able to take her to the school today; she understandably asks; "how am I going to get to school then?" I gave her the same reply that my father would have given me, "Shank's pony." Excitedly Karen said, "Is Daisy May coming up to the Lighthouse?"

Of course I then had to go on and explain the meaning. It is strange how we use colloquialisms and familiarities without thinking that the recipients might not understand, so for those unfamiliar to the phrase …Shank's pony… means to walk

Even one time walking down to Tarbert could have proved enough because the situation was made worse by the preceding events. The day started badly for us, just after breakfast the kitchen window gave up its fight with the wind and shattered, this was seconds after Mags had cleared the table. Still shaking from her ordeal she ran over to the watch room where I was on watch. But for good fortune the situation could have been even worse than I had predicted; but as it was, all I could do was clean up the glass and find something like a board to nail to the outside until it could be repaired properly. Mags had to take Karen down to the pickup point because I could not leave the Station, it was still a very wild and windy morning and the sun had barely risen above the eastern sky, which was of course, if you could actually see the sun through the overcast. It was poor Mags and Karen that had to brave the wind and make the treacherous walk down to the cattle grid. The wind was gusting and more than once it threatened to loosen Mags' grip on Karen's hand and blow one or both of them into the bay, if that wasn't enough; the cattle that roamed free were taking too much of an interest in them and apart from bellowing their disapproval of the wind they began to approach menacingly. Once Mags had overcome that threat and get to the pickup point, both she and Karen would have to find shelter from the biting wind before Mr Shanks arrived. There were only the hedgerows to offer any kind of respite at all and they were both almost chilled to the bone, Mags still had the walk back up to the Station only this time she would be alone and to add to her misery it started to rain.

I was there to meet her, for a while she said nothing then her tears began to fall, "I can't stay here, I just can't." In between sobs and without much imagination I could work out what had happened. There was nothing I could do about the weather but you get wind and rain everywhere and Mags knew this, the danger lie in the location and there really was only two solutions; Mags could stay with Jessie until the situation with Mr Shanks became resolved or put up it until I was able to pass my driving test. At the moment Mags was too upset to consider the options but once she had dried off, warmed up and calmed down, I would put them to her. A cup of hot sweet tea is not the answer to every problem but it does go a long way to mending bridges and does no harm even if as in the case of Sean Connery portraying Major General Urquhart it proves to be a bridge too far. In Margaret's case I hoped that she loved the place as much as I did and this was just a minor glitch; of course I kept those thoughts to myself for fear of setting fire to the bridge and destroying it for good.

This was not the time to take her out the back door to admire the view but by the afternoon both she and the weather had calmed down and by late afternoon the clouds began to break to allow the sun to shed a little warmth on a somewhat troubled Mull.

The visibility improved too and soon we could see clear across to the Isle of Man, Ireland to the West was still shrouded beneath some low cloud and you could barely make out the coastline. On fine clear days when visibility was not hampered by the other extreme of haze; we could see the Mountains of Mourne to the West and to the East as far down the Cumbrian coast as Walney Island, both were the best part of 70 miles away; today we had to settle for the 26 miles to the Isle of Man, but it was more than enough to lift Mags' spirit.

The Mouth of Luce Bay is nearly twenty miles across; at about a third of the distance was a cluster of rocks called the Scares or the Scar, the former is more befitting because any vessel missing the light of the Mull and mistook the Whithorn peninsular for the Mull would have the crap scared out of them when they encountered the rocks.

There's a path that can take you to within fifty feet of the sea and it is still a popular place for rod fishermen; but not one that you would actively encourage tourist to use as the grass slopes are steep and often slippery and a slip or fall could well see you landing into the sea or on the jagged rocks below. I fished there once or twice myself, but my fishing skills were not even rewarded by catching an old boot. I have seen mackerel caught there and one brave lad even caught a sea bass just off the foghorn. Now he was either brave or foolhardy because anywhere other than at the very point where the grassy slope was, the grass ends suddenly leaving at least a hundred feet to the sea below. If you were to catch something that was larger or feistier than you could handle you'd be left with the choice your catch and tackle or your life.

The Mull, because of the vista and the Lighthouse, is a good attraction for visitors and it was not uncommon to find them wandering about the Station and what was worse having them look through your windows while you were eating your meals.

It takes all sorts to make a world and unfortunately there will always be the ignorant, who ignore signs and lack common decency, but generally speaking they were in the minority.

There are some caves in the South West of Scotland that have been attributed as the dwelling place of the legendary Sawney Bean; we had one just up the bay between the Mull and Drummore. To get to it would mean a trek through a farmer's fields and then a traverse down a grassy embankment notorious for adders. Hence the reason why I never visited it, the other being that other than the legend it was just a cave. As for its credentials, well I'm sure that Sawney and his family would have starved to death if they had to hang around waiting on passing traffic for the chance of a meal.

In the late summer we had other visitors to the Mull and they were of the four footed variety. A small herd of roe deer would find some delicacies among the heather and grasses bringing them to the Mull during the evening to feed once the two-footed visitors had left. I had other things to do than make a study of these wonderful animals so I needed a little imagination to work out how they got into the grounds. Was it across the cattle grid or did they leap the six foot wall, it was too much to imagine them being transported by vehicle or the Tardis; if they were capable of one they were surly capable of the other and perhaps they took both options depending what obstacle was in front of them; whatever it was nice to see them at the Mull.

©Steve Hardy, South Rhinns Community Development Trust

Chapter XVI
Fixing the problems

It was now imperative that I take action to resolve the problems that had caused us so much stress, starting with the kitchen window. I telephoned the Superintendent more or less just to inform him of the situation and to confirm that I would need to telephone the Clerk of Works myself. At least he could add his weight if there was any problem getting immediate action. The Clerk of Works was sorry to hear our plight but initially he was not keen to take on board my idea for rectifying and lessening a recurrence. It took my reminding him that the Station works had not been handed over to the Superintendents department and that they had already been notified of the incident; therefore it was highly unlikely that they would accept the works if it was more than likely to happen again. Looking at my proposal again he concluded that it was more than feasible and said he would put it to the contractor to get it sorted out and to expect somebody at the Station later that day.

It was a bit too much to ask for a contractor to arrive on somebody else's promise, but I was not too disappointed when they arrived the following afternoon. I was less than content when their intention was to replace the glass with the same standard glass that had broken and not put in a transom or mullions. Then I told them that if they were going to use the same type glass then they should put in at least one … aesthetically they should make them quarter panes, then the windows would be similar to the rest of the windows at the Station; that after all had stood the test for over 100 years. "That's not what we were told to do," the gaffer said. "Well in that case, I'll just have to phone our Clerk of Works and see what he says." After some pondering and humming and harring, he says, "That means we'll have to go back to the yard and get some wood," "do what you have to do," I replies. It looked very much like the scene from a pantomime, you know the one "there's a hole in my bucket, dear Lisa, dear Lisa" and it looked as if it was going to go on for the same duration, so I stopped him short finally, by saying. "If you think I'm going to spend another day with the wind rattling through my kitchen then think again." This contractor's reputation was let down by poor management and the mentality of the work force. They had the initials R&D and by popular vote it was translated to Rough & Dangerous, I wonder why? I can't blame them for following the Architects drawings but as builders and local ones at that they should have known what the wind was likely to do to the windows. But what gave me greater cause for dismay was what was to follow.

The Clerk of Works duly arrived for a final inspection of the work before passing it on to the Superintendent, well that was the intention; however, he noticed that there was water coming out of a Sulom vent. "Where's that water coming from?" He says, thinking I knew any better than he did. I did have the thought of telling him that I was neither a builder nor was I clairvoyant, but I let it pass. I think I should have worn my uniform or stated my thoughts because he followed it up by saying, "there should not be water coming from there!" just as if I were indeed the contractor. I knew what it was! He had to prove to me somehow that he knew what he was doing especially after the episode with the window. Oh and by the way his comments on that were that it was a great improvement to the safety even if it was less panoramic. It was only stating the obvious so not worthy of a reply.

He metaphorically put on his deerstalker and lit a cigarette instead of a pipe and looked at the offending, drip…drip…drip, "Hmm," he says, scratching his chin; "it must be coming from underneath the house." Now it was impossible for me to resist a little sarcasm, "Elementary my dear Holmes," followed by "I'll go and put the kettle on." I was half expecting a rebuke but he saw the humour in it, especially as it was toned in with the offer of a cuppa.

He had neither the time nor the inclination to start ripping up floorboards in an attempt to locate the leak; for a leak somewhere must be the cause and he concluded that it must be from the bathroom or the kitchen. He entrusted me to keep him informed and to notify him if I had any cause to be concerned about the workmanship. After all it was now passed the date of the hand over and any more problems would delay it even longer.

Once more the workmen arrived in the middle of the afternoon, a plumber arrived and a carpenter with his mate. I began to have serious doubts about this contractor when the plumber left without finding a leak, leaving the carpenter to lift the flooring in the kitchen and the bathroom. Time was getting on and no leak was found, they were about to leave when I said, "Don't leave without at least finding what has caused the leak, or replacing the flooring for our safety," the intention was to do neither, but I said that I would settle with the floor being left open if they could find the leak; at least the plumber would know what to bring with him in the morning. They could see that I was adamant but not unreasonable; and as it was it did not take them long to find the source of the leak. Apparently there used to be a toilet just opposite to where our back door and the tank press is now. Instead of capping off the old redundant piping they just folded it over and it was that old pipe that was leaking under the damp course and out of the Sulom vent. In order to find it they had to cut open the damp course making it useless in the process. Now they knew where it was they could inform the plumber and he could bring with him all he needed to fix the problem. Well that was the course that common sense dictated but not what was to follow. I asked Ken if there were any plans of the Station for the work that was done or for the layout of the pipe work. Now I did my Hercule Poirot bit and got my little gray cells working. I could see the pipe and I could also get the general direction where the pipe was coming from. Looking at the plans I could see that there was a spur off the mains at the back of the house that supplied the old toilet and according to the plans for the renovation, the water for the house was taken from new piping laid at the front of the house. Just to confirm it, I should be able to find a mains cock under an inspection cover outside the wall opposite the engine room and sure enough it was there. Just as I had hoped I could see the spur that led to the toilet but it would be no use turning the stopcock off as it supplied other parts of the station but what it would make easy is to isolate the spur and cap off the pipe. I felt more than chuffed with my bit of detective work but was less than happy when I returned to the Station the next day after a sojourn to the village only to find the plumber in a growing puddle of water trying to cap off the pipe. It was obvious that he did not get my information or he chose to ignore it either way I was a bit exasperated. Once I'd shown him the stopcock and spur it took him less than five minutes to cap it off. I thought that would be the last of the unprofessional qualities of both the company and the workforce, not a bit of it. The plumber was going to leave the damp course without repairing it, I shook my head in disbelief; I says to him, "Isn't that a damp course and doesn't it perform a special function?" He was getting a bit angry at being told what to do and voiced his disapproval; I don't take kindly to threats from anyone and reminded him that his company was under contract and that he was a visitor to the Station if his behaviour did not improve he need not bother coming back to the Station.

Just in case he did not return to finish the job, it was incumbent upon me to inform the Clerk of Works with a full report. Within two hours it was a foreman who arrived and not the plumber, he apologised profusely for all of the trouble I had had and that it was not done to the usual standards of the Company. Now I thought that must be an end to it, why did I even think it; somebody up there has got it in for me. The very next day there was an almighty clatter from the kitchen. I went rushing through and saw Mags in tears nursing her already swollen and bruised brow, pushing the heavy double doors of the tank press off of her.

I have only a basic knowledge of first aid, but knew that there was at least a risk of concussion if not worse so knew that she would need medical attention and not the hot cup of sweet of tea this time. She protested at my attentions at first but was glad of the support I gave her in managing the short distance to the couch. She obeyed my instructions to lie still and I would call the Doctor. Ken was on watch and I was grateful for that, this was getting to be a habit but this time Ken was understandably almost as distressed as I was that Mags was injured. Just as soon as I had called the Doctor, Ken was on the phone to the Secretary, because someone was injured at the Station as a result of an accident or the fault of a piece of equipment in the Boards Charge then the Secretary was always informed. We are not the kind of people make unnecessary claims for compensation but Mags was very fortunate not to suffer anything other than bruising and a headache that lasted a couple of days. However, the Secretary needed to be kept informed during the next few weeks if there was any medical condition that might be attributed to the injury.

Once more I needed to speak to the Clerk of Works; in the six weeks I had been at the Station, I had spoken to him more times than anyone else at HQ in my entire career so far. Anymore and it would risk the gossipers chanting on the grapevine, or worse still Mags questioning my sexual orientation besides my fidelity.

This time he came straight out and asked me if I had a solution. It is OK having a solution after the event but I was still blaming myself for not recognising the danger beforehand. The solution was easy once you saw the problem; the doors were made of laminated chipboard and six foot by 20 inches each they were pretty bloody heavy. The weakest part of the doors; was at the top and bottom edges, and guess where they put the hinges that allowed them to fold together. That one act weekend the doors even more. Once more common sense flew out the window or more sinisterly they might not have followed the specifications, either was an act that might have found somebody liable to prosecution in today's courts had the accident led to a fatality.

We would settle for sliding doors just so long as they were held on suitably strong runners. The carpenter brought along the same hinges with the same type replacement doors to perform the same operation. Not on your life! Finally, the sliding doors performed their job well and what's more you could even remove them singly to allow access to the tank press, without incurring a hernia in the process; much to the liking of the Central Heating Engineer who replaced the whole system at the Station with gas. That was one chapter at the Station I was glad to see come to an end and I could now turn the page.

There was still the problem of the school run to get sorted, what I needed to do was get a car and pass my driving test. Alec had already said he would take me out and give me lessons; that was something I had never solicited and something he asked no reward for, he never even took me up on the offer me lending him the car while I was not using it, for that and being with me for the 1000 plus miles and some good driving tips he

gave me before I took my test I am deeply indebted to him. Just to add to that; he was a first rate Keeper to boot. Alec had a motorbike and while he wasn't on watch, with his family or giving me lessons he was out and about on his bike. More often or not you would find him in the pub in Drummore, sometimes behind the bar sometimes in front.
I paid Archie to take me into Stranraer to have a look at a car I'd seen advertised in the local paper. Not knowing a great deal about cars I reckoned the mars red 1500 cc Golf was great for what I could afford and suited the job admirably. I ignored the salesman's comments about it being a…good bird puller…As the car came with a six month warranty, I was sure that if there were any major faults they would show up within that time, well we can all be forgiven or pay the price for naivety especially when it comes to the small print. On this occasion I was blessed because the car proved to be a gem. Alec came with me a few days later to pick the car up and take me for my first lesson.

It was not strictly my first lesson because I had taken driving lessons when I was seventeen to the point of failing my driving test; nine years later I was about to try again. I had a lot more confidence and maturity and so was comfortable behind the wheel, but I was not complacent. I knew that the more miles I could get under my belt the better my chances of passing my test. They say that if you don't take formal lessons you stand the risk of picking up bad habits from the instructor, well in that case Alec must have been a good instructor and I an intuitive pupil for knowing what was required of me to pass my test.

I had six weeks of lessons before I sat my test; the test centre in Newton Stewart had the easiest of courses and if I could not pass my test there, then I would have trouble passing it anywhere else. We had spent days going round and round the town; so all I had to do was overcome my natural fear of tests. I failed my eleven plus because I was so nervous. It's incredible how the mind works, why should I be so nervous sitting a test yet be so nonchalant when it came to appearing on stage or patrolling the streets of Belfast for that matter? I needed to pass my test but that need may also put undue pressure on me and make me worse; I had to find some way of neutralising my fear.

Since my Supernumerary days I had become an ardent reader, one of my favourite reads was the Readers Digest, I had come across an article that shed light on the very subject of fears and phobias and there was one case that did mention Test Shyness as they called it, the person overcame their problem by undertaking a task that would release a surge of adrenalin; the adrenalin in the system stayed with them long enough so that they were already on a high when they came to sit the test and as a result felt calmer in their being as they came down from that high. The person in question went skydiving for the first time; there was no way I was going to go skydiving even had there been the facilities to do it and I was paid a million pounds to do so; and I was not going to go bungee jumping either. But my respect of heights would give me enough of a rush to do the job just nicely. It just so happened that my test day coincided with the weekly shopping trip so Mags would be away from the Station when I made the decision to paint the dome. Ken thought I was mad but Mags would have gone ballistic had she known what I was about to do.

He may have thought I was mad but Ken showed little signs of it when he supplied me with the paint, brush and checked my harness before I scaled the stairs. The day was bright and sunny and the wind just a gentle breeze, an ideal day for painting the dome. It was not on the list yet for painting so I stood the risk painting the dome for nothing, if I fell or I failed my driving test then yes I would have painted the dome for nothing. Weighing up the risks I thought it well worth it, besides it would not be the first time I had painted a dome.

It was however the first time that Mags was to see me on the top of the dome and that is what happened when she arrived home earlier than expected. I was within a couple of feet and about twenty minutes of finishing when the car appeared round the corner. I would have to suffer the indignity of being treated like an errant schoolboy until she calmed down enough to realise I was safe and the job was done. I loved her for it nonetheless.

Apart from stalling the car at a junction and recovering from it like an experienced driver, the test was a doddle. I had expected the test to last at least a half hour, so when we returned to the centre after just twenty minutes I knew that I had either passed or failed. If I was that bad the examiner would have asked me to stop the test on the course. I was delighted that once I had answered his questions on the high-way-code, the test was over. I was ecstatic when I was passed and was given my certificate.

I returned to the Station without my L-plates to the congratulations of Ken and the family who had sent Karen up the hill to act as scout awaiting my arrival. The hill was the highest point of land on the station and its elevation hid the entrance gateway and the cattle grid from the Station. From the top of the hill you could get a good view all the way down to the Cattle grid at Tarbert. On the hill were the flagpole and the fresh water tanks for the Station it was also a favourite place for the kids to play.

We now had the means to explore the whole region as well go into Stranraer and visit Jessie when we had the time. More importantly I would have the means to visit Mags in hospital when she went into labour; Mags was by now two months pregnant and the baby due in the middle of December.

©Steve Hardy, South Rhinns Community Development Trust

Chapter XVII
New Arrivals at the Station

Tats had got used to the idea of being the only pet on the Station; he was the most independent creature I had ever come across; but his affection for us meant that he always came back, even after a few days of sowing his oats among the cat population from the Station to Drummore and beyond. Sometimes he would spend a few hours with the Keeper on watch whoever it was and then he would disappear, all the Keepers had a soft spot for Tats; even if he had the bad habit of walking over the Met book just as you were reading out the figures to the guys at Prestwick Airport or worse still frightening the crap out of you as he silently walked passed with his tail erect when you were checking the other lights at midnight. At least he did not have the reputation of the cat kept in the local store in Drummore who was famous with the ladies for passing between their legs with his tail in the air, much to the delight of the men-folk there, when the sudden and unexpected cry of, "Ooooh" erupted, to announce to everyone that the cat was back and on his way to the house at the back of the store.

Tats had a much cleverer trick up his sleeve; for some time we had been puzzling how he got in the house in the mornings, we assumed that one of the keepers had let him in but had left the door open. We did not mind about him coming in but we did not like the idea of the door being left open, so after a while I thought I'd mention it to Jimmy McGowan when I changed watch with him at 6 am; Tats was not home yet but he was usually in the house by the time we got up at seven.

Jimmy was a little amused when I questioned him and right on cue Tats appeared at our door; "watch this", Jimmy said. Tats jumped up at the door and clung to the handle. Unlike most doors our door opened out the way, what's more the handle was one of the round polished brass variety and tended to be stiff at the best of times, so much so that even Karen had trouble opening it. What I was about to see was literally an amazing feat of strength, agility and perseverance. Tats swung on the door from left to right again and again until he heard the catch; he then did a twisting dismount from the handle that opened the door a few inches in the process and walked proudly in. After changing watch with Jimmy, I went back to the house to give Tats something a little special as his reward for treating me to something equally special. Try as I might though I could not teach him how to close the door behind him.

I thought my days at the Mull were about to come to an abrupt end when Ken and Mrs Clark had brought a replacement for Tiny and Darragh to the Station; once again she was an adorable little Chihuahua and because she was just a pup, the tiniest dog I had ever seen. They named her Daisy and she was not much larger than the head of a sunflower that I used to think were just giant ones. She was kept indoors for the first couple of weeks but once the better weather came, Ken and Mrs Clark took her out onto the green. Heaven knows what Tats must have thought when he spotted Daisy for the first time but by the way he was stalking her it looked very much like he was about to catch his dinner. Lucky for us he was within easy reach so I swiftly gathered him up and took him into the house. He really was a lovely cat and did not mind a bit for being distracted from an easy fresh meat meal; we would find out later that he was a very accomplished hunter and that would make this episode even more remarkable. I would like to think that Tats had a sense of humour and did it just to scare the shit out of us.

Of course with Mags expecting we knew that there would be another new arrival at the Station come the end of the year. What none of us knew was that we would be getting a new arrival of the four-legged kind ourselves and this time it would be all down to me. At the end of June that year with the schools breaking up for the summer holidays, we thought Karen might like to spend some of her holiday with Jessie. I would take her up to Patna myself while Mags stayed at home, the thinking behind it was that once we got to Patna either or both of them would start greeting when came to say goodbye and Karen might end up coming back with us.

You may be thinking that there must have been something wrong with the relationship of father and daughter, not a bit of it; Karen loves me dearly but it was easier to part from me because she was used to it what with me going to the Rock

We used to attend Karen's Parents night at Pirnie Hall and her teacher would always know when I was out at the Rock and when I was coming home because of Karen's work. It would fall away for a few days after I had gone out to the Rock and reach a peak a few days before I came home. Maybe that was an influence on my deciding to put in for a transfer.

After I had dropped Karen off I would take my father in law fishing up at Loch Doon, now that I had the car I would not be getting sore feet; I did not get any fish either, neither did Jimmy; what we got instead was eaten alive by midges. Did you know that it's only the female of the species that needs blood, now why was I not surprised on learning that gem?

When we got back to Burnton it was still light but I was tired enough just to want to have something to eat and go to bed. Earlier in the day I had been introduced to the last of Jimmy's dog Sarah's pups, the other four had been given away to locals. I could not help but wonder what was wrong with this one because she was a good looking pup even if her lineage could not be traced; perhaps her only fault was that she was the only bitch in the litter.

Jimmy kept her shut up at night in a coal shed, don't be so harsh, the coal shed was inside the house and at the back was the hot water tank so it was warm. The trouble was that the door had seen more than one litter of pups and the bottom was chewed away in parts. During the night I was awakened by the gentle nudge of a muzzle on my hand followed closely by a whimper. You know I really must be a sucker for the doe eyes of a female so how could I ignore the pleas of this young lady. Like our first night together, she would spend the next seventeen years sleeping beside us.

It's the things we do on impulse that can create the biggest problems, yet we tend to forget about them at the time. For instance, what if Tats didn't get on with Kelly? Well for all the people out there who do not keep pets; I am going to tell you something that you may find amazing if not unbelievable. From the minute Tats and Kelly met each other, it was like love at first sight. Tats simply adored her and like the male that he was he took the dominant role of the relationship. My understanding was that cats were territorial, that being so you would wonder why he would tolerate any other creature on his patch let alone show true affection for a dog. That is one of the mysteries of the pet world that dispels some of the myths. First of all there was the relationship between Tats and the hamster, now there was the relationship between Tats and Kelly.

You may think that it was just that Tats was an unusual cat, or maybe it's that we are unusual people because seventeen years and a menagerie later we have got two dogs a cat and four goldfish. All of them get on well, and that includes the cat and the goldfish, although I was troubled one Christmas when Mags bought the cat what I mistook for a fishing rod. Well all I could see in the wrapping paper was this wooden stick and a bit of string, what else was I supposed to think? Can I dispel another myth, you know the one about goldfish not being intelligent or having the will or ability to interact with humans.

Well we have one goldfish called Pearl who attracts our attention by spitting high-pressure water at the side of the tank and then performs a dance to show us that she wants food or attention. I think she must be a bit upset with us because since we moved the fish tank out of the living room she does not perform so often. Now I may be quite wrong when I say she after all if you are not an expert how can you tell what sex a fish is if you don't see it laying eggs. Enough of a divergence; let me transport you back to the Mull and the summer of 1982.

Tats was usually the instigator of the mischief between them and sometimes it was like having another couple of kids in the house. It would usually start when Tats had woken up from his afternoon siesta. Tats' pride of place was on his throne…a cushion…on the settee. Kelly had to relinquish her place there whenever Tats came in and she was relegated to the floor below him. He would look over the edge of the settee to see if Kelly was sleeping, he would purposely wait for her eyes to close before reaching down with his long feline legs and tap her several times on the head with his paw; claws sheathed. The pair of them would dash about the house if the front door was closed or run out onto the green if it was open. Bundling around the house sounded like the same herd of elephants as when the girls were carrying on; and so the pets got the same treatment. We would open the door and tell them to do their carrying on outside.

There will probably be more than just a few photographs in the possession of visitors who were enthralled and entertained by the spectacle of Tats and Kelly play fighting.

I would dearly have loved to film one episode and be £250 richer from its exhibition on …You've Been Framed; that was when we were taking Kelly for a walk and Tats followed us. A black Labrador was cresting the hill when it spotted Kelly, you cannot be sure what the dog's intentions were, but Tats was not going to take any chances. He being the dominant partner felt honour bound to protect Kelly by making himself as large as he could, arching his back he made a sideways attacking sortie towards the dog, all the while hissing and spitting venomously.

All the bravado of the Labrador evaporated in an instant and he turned with his tail tucked between his legs back the way he came, yelping as if the hounds of hell were on his heels; passing his owners in the process and not stopping until he felt safe. Heaven knows what the owners must have thought when he passed them, but they did not make enquiries of us even when we appeared with Kelly. As for Tats, he was taking another route keeping a low profile through the heather because something else had caught his attention. They must have thought Kelly had made to attack their dog but on finding that he was unmarked, chose not to remonstrate with us. We would have been more than delighted to inform them that the only thing hurt about their dog was his pride.

Tats had a comical relationship with the cattle especially if they were close to the cattle grid at the entrance to the Station. For some reason the cattle were accustomed to dogs and were only wary of them if they were not on a lead. They had no fear of dogs and would posture and gesture in feint attack to ward off any. They would do the same with Tats but he would taunt them by hiding under the bars of the cattle grid and popping his head up, then down again if they got brave enough to ward him off. He would do this several times and when the cattle were brave enough and close enough he would arch his back and go on the attack scattering the cattle in the process. With Tats with us we could all enjoy our walk in peace and the cattle would be left to graze somewhere else; with at least one pair of eyes open for Tats.

There must have been more to Tats that went unseen; it's a well known fact that domestic cats like their wild cousins do not like carrion and will only eat it in extreme circumstances, they would much prefer to eat fresh meat caught by themselves or out of the tins provided by their owners. Tats was no different, during the early hours he must have caught a hare, and I assumed it was a hare because there was not much left of the carcass. The giveaway was the length of the hind legs and the general size of the body. Tats was determined that Kelly should share in his catch so he brought her home a hind leg with still a bit of meat on it. Not the nicest of things to have sitting on your living room floor, Mags was definitely not amused and insisted I throw it away. Well throw it away I did, three times in fact, every time I threw it away Tats found it and brought it back. It would have been no use throwing it in the bin, if Tats could open the front door; then he would have made light work of a bin. In the end I threw it over the cliff making sure that I landed in the sea; even Tats would not overcome his natural aversion to water to retrieve it. Tats enjoyed his time at the Mull as much as we did, but events sometimes come along that can tear apart the bonds of love; and with a cat as independent in nature as Tats it was to take the form of the arrival of a new baby in the household. It's easy to attribute our pets with human feelings but what really went on in his mind we will never know; maybe he thought that he would not be loved so much with a new baby needing constant attention. I would like to think that Tats thought the rightful place for a dog to be was beside his/her master and that Kelly would be the ideal companion for the baby boy as he was growing up. Whatever, a few short weeks after the arrival of Gavin at the Station, Tats disappeared, never to return. We thought we caught sight of him once or twice during the years to come but nothing definite, knowing Tats he must be sire to a score of cats and it was his likeness we saw in them. I do believe that his spirit is wondering the Mull to this day even if his body is not.

We expected Gavin to make his appearance around the 12th December but Gavin had other ideas. Dr Hocken was not as sure as we were on the sex of the baby Mags was carrying. He did not share in our intuitiveness and suggested the size of Margaret's ever growing belly was merely due to an extra amount of water she was carrying. Gavin was our third child and as there were no complications with the first two and this pregnancy seemed to be going the same way, Mags was only given an ultra sound during the first three months. Nearly a week passed and even Dr Hocken was becoming concerned. We'll give Dr Hocken the benefit of the doubt; that his concern was genuinely for the health of both mother and baby and God forbid that it had anything to do with Christmas fast approaching. He decided that Margaret should be admitted to the hospital in Stranraer in the hope that she might be delivered over the weekend. I had taken a two weeks paternity leave so that I could be present at the birth it now looked as though Gavin had other thoughts and wanted to share Karen's Christmas Day birthday with her. Dr Hocken was adamant though; if she was not delivered naturally over the weekend she would go to The Creswell Hospital in Dumfries on the Monday to be induced.

We were the only family living at the Station at the time, Alec and his family had been transferred and Ken had bought a house in Stoneykirk called Hunters Moon; he bought it as a retirement home for when he retired in a couple of years. He commuted to and from Stoneykirk and if he was on watch again at six in the morning, he would sleep in the caravan provided for the Local Assistant to do the same.

During the summer the tower was sand blasted and re-painted, the houses were also undergoing renovation, so just as in our case it was better to do this while it was empty. We saw more of Jimmy Fyffe and Jimmy McGowan at this time than at any other period. They would have to cover for Alec who had been transferred, for the rest of us on our days off and for me while I was on paternity leave. The girls had already grown very fond of Jimmy Fyffe and they would spend hours in the Bothy tormenting him or keeping him company. Jimmy was like a member of the family to them and we entrusted them to his safe keeping; now that it became obvious that Gavin would not arrive until the last day of my paternity leave; I had the un-envious task of asking Jimmy to hold the fort till I got back from Dumfries. Keeping an eye on the girls was second nature to him, and once he was assured that they were safely tucked up in bed, he would pop over every so often to check on them.

Both Jimmy and I knew that it could not be guaranteed that I would be back for my watch at 2 am on Wednesday the 22nd December, but Jimmy was prepared to stay on watch till I was able to relieve him.

I arrived at the hospital at about 11.30, Mags was just getting prepared by the nurse to be induced. It was going to be a long day for me so heaven knows how Mags was feeling. I was present when Karen was born and could only imagine the pain she was going through; I was there when Kirsty was born, but only just because unbeknown to me Margaret had spent most of the day going through labour before calling for me to come home from work and sending for an ambulance. She said that giving birth to Kirsty was like shelling peas in comparison; by the pain she was having now this delivery was going to be a pumpkin. The Doctors were growing concerned that both Mags and Gavin were becoming distressed so at six forty five the decision was made to send them to theatre for an emergency caesarean if need be. Gavin had had enough and made decisions of his own, he still needed a little help getting his broad shoulders out but other than that he was a fine healthy eight pound three ounces and he arrived at seven thirty. Once he was cleaned up he was passed to me while they were attending to Mags. After a little whimper Gavin settled quietly in my arms opening his eyes, I'm sure I could feel him say "Oh Daddy, thank God that's over," to which I silently replied, "Yes Son, thank God." The midwife seemed certain that Mags needed rest more than anything else, so began to dress Gavin ready to put him in the crib. The thought of it rejuvenated Mags into action, demanding the midwife pass Gavin over, she rebuked me with her eyes without needing to say anything. I have had some proud moments in my life but none more than being present at the birth of my three children, watching them develop into fine young people and achieving the goals that they have set in their lives. What made those moments possible though was the love and devotion that Mags has as both a mother and wife.

I got back at twenty five past twelve but shattered after the day and the ninety-mile journey back to the Mull. The girls would never have forgiven me if I did not wake to tell them that they had a brother, they were a little disappointed that I had not brought him with me, but I assured them that he would be home with Mum in few days.

That few days turned into a just over a week because the weather had turned bad and there was snow on the ground and the roads treacherous. The girls thought I was lying because there was no snow at the Mull. That was the strange thing about the Mull; it had a climate all its own. Sometimes we had balmy sunshine while the rest of the south west of Scotland had rain, if it was not for the extreme of wind we could have grown Palm trees like those at the Point of Ayre lighthouse across the way on the Isle of Man; or some of the exotic plants that were grown in the Botanical gardens at Port Logan just up the Coast.

We had to spend that Christmas without Mags, to make up for it I made everything for us to have a party but the smiles on our faces were just for show and the camera, it was not the same without Mags and the birthday present of Gavin that Karen claimed as hers.

©Steve Hardy, South Rhinns Community Development Trust
Mull of Galloway as it is today with its visitor centre, museum and the Gallie Craig Coffee House, as Scotland's Most Southerly Point it still attracts thousands of visitors each year

Chapter XVIII
Sprucing up the Station

The Tower was looking good after its new coat of paint; however, there was a problem that was made known to us after the first heavy downpour of rain. The grit blasting had weakened the mortar and the joints became porous in places. All the good work had to be laid bare again and the tower re-pointed and painted. All the grit blasting had left deposits of grit around the Station; and it took weeks of cleaning before the Station was free of the stuff.

It would not have taken a genius to work out that with all the expensive renovations going on at the Station, the Commissioners would surely want to pay us a visit on their next tour. It took no more than a phone call from the Superintendent to determine what needed to be done as far as the painting was concerned, in short everything that was not covered by the contractors in the renovation. The Superintendent would visit the

Station once we had finished the painting and about a month before the Commissioners set sail.

Just to make sure that there were enough hands to do all the work, Ron Ireland and his family were transferred to replace Alec Dorricot. Ron would take over as First Assistant, I did not mind one bit, Mags and I were just happy that the Board had sent a family here who were much the same age as our own. Now there would be fewer problems with the school runs.

Ron was another Keeper with the Service in his blood, his father Ron senior was Principal Keeper at Scurdieness Lighthouse and was due to retire shortly. His wife Anne is the daughter of one of the Stewards onboard the Fingal. And they had three children; Gary, David and Jacqueline. Ron was like me when I first came to the Mull; he did not drive, but was going to make it a priority. Ron was affable and easy going and the whole Station seemed to meld into a good working unit. Ken needed to give little or no direction as to what needed to be done, he could relax and continue his commute to and from Hunter's Moon; only now he had the house at the Station and not the caravan to sleep in. We only saw Mrs Clark on the rare weekends that she came to the Station with Daisy.

Ron did not take me up on my offer of giving him driving lessons and chose to take professional tuition. After he passed his test he bought an Austin Maxi; and we would take it in turns taking the children to Tarbert.

The weekly shopping trip needed to be sorted out because there simply was not room for all of us in Archie's car. There was another problem; Archie was nearly 80 and his vision was deteriorating, we no longer felt safe with him at the wheel. It was a good time for change and to relieve Archie of the burden. The solution was easy and right before us; Mr Shanks used a mini bus for the school run and the Board had already made use of his service; so need not put the run out to tender, as would be the normal case; they would just extend the service.

One of the worst jobs and dirtiest was tar and chipping the roadway. We had to do this before we started the emulsioning and before the hotter weather made the task unbearable. It was not like the tarring that you see at road works. There the tar and the chips come ready mixed and heated before being laid on the surface to be rolled and let to cool and set. Our tar came cold in 45-gallon drums and was the consistency of maple syrup. The barrels had to be given a good roll beforehand just to mix the contents up a bit to try and make it consistent. It all depended on how old the barrels were and how long they had been standing on their ends, you could tell because one end was heavier than the other where the tar had settled. With cases like that you were knackered before you even started on the road, with end over end rolling.

We were lucky to have Jimmy with us because of his farming connections he could get the use of a tractor and trailer. We could load the barrels onto the trailer instead of humping them around on the barrel barrow. The only thing we needed to cart around were the chips that had been delivered by lorry. We kept the special watering cans, rakes and shovels in a store next to the garages. We rarely used the garages for the cars; it was much easier and lazier to park the cars on the gravel of the courtyard.

The kids used the garages as a playground where they could imagine them to be whatever they chose. Before they were garages, they were stables for the horse and cart, and if somebody had left one of the doors open they could climb up into the loft. The back of the garages was lower than the front and almost backed into the hillside, the children used to climb onto the roof and slide down; landing on the heather to cushion the fall; but only if there were no Keepers about to see them. Apart from that incident of silliness we trusted the children not to play outside the Station near the cliffs, that trust was never betrayed by any of them. There was never as much as one incident where the children had to be warned.

I had to use the oldest clothes that I could find to do the tarring, and at the end of the day they would have to be thrown away as well as my Wellington boots. It always amazed Mags how I could get tar down my boot and yet not cover my socks with the stuff. The answer to that one was simple; my socks had long since rolled off my feet and ended up in the toes of my boots. The object was to sprinkle a layer of tar across the road then scatter chips over the tar in a single even layer. The weather would play an important part in deciding when to do it. It had to be a dry day, that goes without saying, but it could not be a scorcher otherwise the tar would not set and the chips would be scattered ever where but on the road. We had to have the job finished at least a couple of months before the Commissioners were due. The tar should have set well by then yet still retain enough chips to have the appearance of being recently done. We still had all the painting to do as well, so basically we were left with about a three-week window of opportunity. The weather was on our side around the middle of April even though we had some ground frost to contend with at the start, it would not matter once the job was finished because it would help the tar set, but you did not want the tar to set before the chips were scattered. It was a job that thankfully we could accomplish in about 5 hours thanks to Jimmy and the tractor, and once it was done we could announce its completion with the ceremonial burning of the tar laden clothing in the garden incinerator. We would notify the coastguard of our intentions just so that they were aware. It would not be the first time that a similar event had caused the coastguard to be called out only to find the Keepers having a barbeque.

Guy Fawkes Night was interesting at the Mull or not so interesting if you compare it to the way others were allowed to celebrate it. It was OK for us to have a rip-roaring fire in the Stations garden but not to set off fireworks especially rockets; they could easily be mistaken for Distress flares. It did not matter to our children because I had seen the effects of the misuse of fireworks in life and not the less graphical images portrayed in the government's television ad. The girls were old enough to make up their own minds about the subject and if it was important to them we would take them to an organised display, as it was they settled for the much less dangerous sparklers. I think Ron's children were the same; but we did treat them once when it was time to renew the rocket flares that we held on the Station. The flares were used to answer a distress flare, they would send up a distress flare and we would reply with three. It was a rare event and one I was to see only once in my thirteen years in the Service.

The flares would be issued to us by the coastguard and they would notify us when they were due to be renewed, generally they would uplift the old ones; but on this occasion we were given permission to set them off on the pretext that it was a training exercise. It was not a fib to say that we had never had the experience of setting off a flare, but on the other hand we could all read and we were not stupid, so the coastguard would know that there was an alternative motive. We had six flares on the Station and compared to the multicoloured and explosive genius of the display rockets ours were a bit of a damp squib but the kids were amused and the bigger kids even more so.

The emulsioning of the Station was pretty straight forward unless you lived in number three lighthouse cottages, where coincidentally I lived. At the back of our house the boundary wall continued on for another eleven feet after the cliff top had disappeared to the nothingness of a two hundred and thirty foot drop; it was the same on the opposite angle. I was lucky in one respect; they'd had the foresight to build a brick shed at that corner and its walls acted as the boundary, where as the rest of the Station boundary was built dry-stone fashion with the surface uneven, this corner was relatively smooth so I could use a roller on an extension to paint it. It was still a job that needed concentration because I was quite close to the drop; so you can imagine the fright I got when a visitor interrupted my concentration. We do not usually get visitors roaming around this side of the boundary wall and especially not so close to the edge. Had it not been for his question, announcing his presence I think he would have got a face full of paint as I turned round with my pole and brush in hand. "Are those penguins down there?" Just in case I had been transported to the southern hemisphere, I looked to see what he was pointing at. Sitting on an outcrop of rocks just above the high water mark were a gathering of razorbills and guillemots, I could understand his mistake especially as his South African accent showed his origins. His question and his accent had diffused what could have turned into a more confrontational meeting. As it was; I still had this chap's safety on my mind as well as concerns that he might not be alone. Seeing no one else I answered him, "No they're guillemots and the ones without the white are razorbills." Just at that moment one took off, if he had watched and waited for that moment he would have known they were not penguins unless they have a variety down there capable of flight; but then he would have got the full force of a couple of choice profanities for scaring the crap out of me with his sudden and inappropriate introduction, I did however remind him politely of the dangers of going to near the cliff edge. I needed to get this wall finished so did not have time to talk too long with this well travelled visitor, he would be one of many visitors to the Station that would not be able to go up the tower in the afternoons. We simply had too much work to do until the Commissioners had made their visit. In the end we put a sign up covering the normal one that said… Station closed until further Notice…

We had now painted everything at the Station, with the old army dictum…If it don't move paint it…god help the kids if they stood still for too long; not much chance of that with our lot. They had the sense not to play near the cliff and to stay away from surfaces that had just been painted, the same could not be said of Kelly; there was more than one occasion that she came home with some additional colours to her coat. If she had been a dog rather that a bitch we could have called her Jacob, in any event she seemed happy for the extra bit of fuss we made of her. They say that dogs can pick up a lot of human characteristics; I believe that to be true because I would swear under oath that Kelly had a sense of humour and tried desperately hard to smile. If a dog bears its teeth, it is said to be a defence mechanism or a sign of aggression, but our Kelly would raise her lip on the one side of her mouth with her head bowed a little and she did this within context of something humorous just happening that she had been involved in.

There was one place outside the Station that was still under the Board's jurisdiction and that was the small pier and shed at East Tarbert, it had not seen use since the extension of the single-track road all the way up to the Mull. It still had to be maintained; the shed had to be checked to see if it was still wind and watertight and the paintwork freshened every few years. The pier would need to be cleaned with oxalic acid to kill and remove the build up of sea growth that made the surface slippery and dangerous to walk on. The Superintendent had paid his visit and this was the only area that needed any work doing to it before the Commissioners came; at the moment neither the date of their visit nor their mode of transport to the Station had been confirmed but the grapevine had it that they were on the West Coast and should arrive in our neck of the woods within the week. The Superintendent did relate the possible sale of the land and buildings down at the pier as the Board was doing much the same with all of the excess land in their possession, so the sprucing up of the pier might serve two purposes.

The Pharos berthed at Cairn Ryan at the old jetty used for dismantling Warships; I remember seeing the aircraft carrier Bulwark getting dismantled on a previous visit to Stranraer. The jetty was only a little way from the Lighthouse called Loch Ryan; it used to be a one man Station but was made automatic in the sixties. The Commissioners had hired an executive minibus to take them round the Lights of the Rhinns, starting with Corsewall Point, then going on to Killantringan before finishing up at the Mull.

This would be the first time that Mags had seen the Commissioners and the excitement generated were second only to that of a Royal visit. They were the people responsible for seeing that we remained in our jobs and that the money spent within the service had value. For the children it must have been a big letdown not to see anyone they recognised from the Royal family. The Commissioners split up into groups after the main inspection tour with two or three invited into each of the houses for tea. Mrs Clark was well known to them and treated in the same way as you would a long lost friend, Mags was her usual shy and timid self when it came to distinguished visitors. Apart from general chitchat one topic that went unmissed was the problems that we'd had and their interest in seeing that everything was OK now.

Sometimes the Commissioners let slip things that would later appear in the Lighthouse

Journal; it started off as a question just to see if I was clued up on the history of the service, " Do you know it's our Bi-Centenary in a couple of years time"? There was not a Keeper worth his salt did not know that the NLB was formed in 1786, whether they could do the math or not was an entirely different matter. I knew I was about to be told something important just as I knew that I need not have to go into detail my knowledge of the Service, so I just said, "Yes." He continued, "Well there might be something in the offing in order to celebrate it, we're not too sure yet but we think it might be a book." It was said in the same way as you might expect someone giving top-secret information, for your ears only type of thing. I expect he had said the same to all the rest of the keepers he'd met on this tour and as there were only the Keepers on the Isle of Man to go; we were among the last.

The visit concluded with the Commissioners going down to East Tarbert just as the Superintendent envisaged. I was on watch that evening so had to keep my eyes open for the Pharos to dip to her as she went passed.

Number 3 Lighthouse Cottage's back yard with the Irish Sea 230ft below

Chapter XIX
Striking Oil and The new Central Heating

Gavin was growing up fast and from the moment he could walk he became quite a handful; he was no longer the baby of the Station because just like Mags, Anne became pregnant within weeks of arriving at the Mull. It must be something to do with the air. Gavin was very much an athletic toddler and thought nothing of gate vaulting over the side of his cot; he would use the padding of his nappy to cushion his bum on landing. More than a few times we woke to find the front door open and Gavin sitting on the step playing with the gravel, the first time it happened we thought Tats had come home, but our faces dropped on finding Gavin out in the courtyard.

In the end we would have worried ourselves sick so we had to give Gavin the same trust that we did the girls; Gavin was a contented baby but now that contentment had curiosity thrown into the mix we had more to worry about than his wandering about the station. As he got older his curiosity would get him into mischief, he was aided and abetted in his crimes by David. It would be hard to lay the blame on either of them, as they both seemed to have the same mischievous streak.

Many a time the Keeper on watch would find the rain gauge full when there had not been any rain, and both Ron and I had found our tyre let down, that was about as serious as their mischief went and the tyre incident only happened once.

We never found out who it was that went into the oil store and opened up the tap to a 90-gallon oil tank but it took the best part of a day to clean up. Luckily only a quarter of its contents were spilled; even so, oil is not the best substance to clean up without the threat of pollution. We needed to find a 45-gallon drum to put the oil we could scoop up with a shovel, first we needed to cut off the lid with a hammer and chisel.

Quite a bit of oil had escaped from the store and had run onto the courtyard onto the gravel, the only thing we could do with that, was to lift the oil soaked chips and much of the saturated soil as we could. Then replace the soil and rake over some chips. The hardest thing was to dispose of the drum of oil and waste, luckily we had Jimmy Fyffe who got loan of the tractor, the drum was taken down to the farm to be uplifted with the other waste for specialised disposal.

Coincidentally around the same time we had notification from the Board that an exploration group were interested in setting up a transponder on our balcony to help with geological mapping of the area. There had already been discoveries of natural gas out in Morecambe Bay and oil deposits in other parts of the west coast; those discoveries were largely due to the use of satellites. Perhaps the spillage was worse than we thought; and had been picked up by an orbiting satellite; Nah, it had to be coincidental, nevertheless it was a yarn we could tell the children especially the culprit/s who opened the tap just to emphasise the seriousness of it.

The chaps duly arrived to set up their transponder, they did not have to survey the area to pinpoint its exact location; that job was already done for them. The latitude and longitude were marked on any marine chart and there was a trig mark on the courtyard wall indicating elevation as well as position information that could be found on an ordnance survey map. To save them having to come back on a regular basis to charge the batteries, we did it for them. They were over the moon at the offer, for us it seemed like nothing but a matter of turning a switch; there was a bit more to it than that but nothing to warrant the amount of thanks we got both before and after the event. We had a personal visit from one of their head people and a bottle of good single malt whisky each as a measure of their gratitude.

Well you already know how I feel about whisky especially after the incident with Bob Duthie; my bottle looked destined to remain in the cupboard for a rainy day or for a rare visitor to our humble home. I'm not sure if it was raining the day that Karen got chicken pox but I was pretty glad that we had the whisky on hand. Karen had taken the chicken pox worse than the rest, she had spots in her throat and her glands were swollen. It seemed as though there was nothing that would help relieve the irritation, I already knew about the anaesthetic qualities of alcohol as well as its medicinal values, whether it was right to give Karen any was debatable, but its effect was nothing less than spontaneous. Karen said it was like velvet sliding down her throat soothing as it went. Both Mags and I reckoned that she had been watching too much television and had picked up the saying from there; even if just days before Jimmy Fyffe had set a mole trap outside the gate and caught a mole and allowed the kids to touch the fineness of its coat, I could not imagine him dissecting it for them to swallow.

We had to keep our eyes on the bottle though; just in case Karen took the same notion as she did with the syrup of figs. Talking about children being under the affluence of incahol, I am reminded of the time that Ron was making homemade brew in the washhouse at the side of the Principal Keepers house. Ron had inadvertently left the door unlocked and Gavin and Jacqueline had got into it, the brew was proving in the bottles and nearly ready for consumption; had they managed to undo the stoppers then the content would have been powerful enough to inebriate two small children. We could just imagine having to call the doctor out to give them something for the skitter, dehydration and a hangover. We would have thought that Gavin would have learned his lesson about curiosity and experimentation; not on your life, there would be quite a few times we would have to take Gavin to the Doctors or phone the help line because he had taken something that might be toxic. Over the years he has controlled and channelled his curiosity and need for understanding by graduating from St Andrews University with a BSc Hons in Chemistry, this year of 2009 he graduates with a PhD.

Racal Decca were also interested in the Mull as the location for one of their new systems. They installed the system on the balcony and had a monitor in the tower. They also sent a chap over to monitor the system and he stayed in a caravan and kept in touch with his HQ by radio. He came from Northern Ireland and was with us for about four months, he found it lonely without his family; his wife could not come over because she was expecting their first baby. She did eventually come over to spend the final weeks of his stay.

About the same time The Mull was designated an SSSI...Site of Special Scientific Interest...So it looked as though the Mull of Galloway was going to increase its popularity in ever abounding steps. Geographically the Mull of Galloway is the most Southerly Point in Scotland; it's no coincidence that there are Lighthouses at all the four cardinal extremes. Girdleness on the East Coast, Dunnet Head on the North Coast, Ardnamuchan on the West and Mull of Galloway on the South.

N° 3 Lighthouse Cottages Mull of Galloway was the last house in Scotland...I would dispute with anybody the right for it to be claimed the reverse as the 1st house in Scotland, simply for the fact for anyone to get to it they would have to anchor their vessel on an unforgiving coastline and scale a two-hundred and thirty foot cliff. That honour should go to the house that has access by road from north and south of the border. The Mull was already very popular with tourists but it had not got to the stage where there were bus tours; the single track was simply not a road that could handle that kind of traffic even if the buses could find somewhere they could park and turn around. Even still there were enough visitors by car to warrant the Council being asked by the farmer to make a car park for the ever-growing number of cars parked just outside the station. We could well understand his dilemma because had they ignored the signs and came into the Station with their cars then we would have no alternative but to put a gate on the outer boundary; it must have been so frustrating for him to see the ground being torn up by traffic. It was land that was already lost to him but the more the ground churned up the more the vehicles would encroach on other ground. If the Council did not make a car park then there would be major problems with access. The road was listed as a thoroughfare and the Lighthouse had to have open access but the Farmer was entitled to have his land protected, the tourists were the culprits but the area needed the tourists to help with the economy a vicious circle that could only be sorted out with the council forking out for a car park.

We were still bothered by the odd visitor ignoring the signs and arrogantly parking their cars on the roadway by the garages, they were the type of people who would argue their right to park anywhere and vehemently express that right with the ignorance of saying that as we were a government body it was his taxes that paid for our upkeep. It mattered to them not that our Service as with Trinity House and The Commissioners for Irish Lights had their funds generated and paid for by the Ship-ping Companies whose vessels passed the lights and paid a certain amount for the service. The only persuasion that had any result was that if they did not move the car we would inform the police of the obstruction caused; if the police chose to apprehend the vehicle on its way back and lecture the miscreants on their unacceptable behaviour then that would be the chance they took plus they might also incur other penalties should their vehicle not meet the road safety requirements. The most annoying part was that if they had asked we might have let them park somewhere where they were not going to be doing any harm.

We often attracted events like cycle races from Stranraer to the Mull and Vintage Car Rallies. Now that really was a spectacle when all these old cars made it up the steep hill, well nearly all of them, they were the only cars that we let into the Station en mass. One was of particular interest because it was not an internal combustion engine but a steam car. Not only could we fill up his water tank but give him some paraffin as well, we only used paraffin at the Station for cleaning brushes and the like and the small amount he needed would barely be missed.

Summertime at the Mull was idyllic and the time when most visitors envied our lifestyle and location; compared to many of our brethren…poor choice of words there; that's what they call the Board of Trinity House… Our fellow Keepers throughout the world who are more isolated and face the rigors of a much harsher climate, than when God is in his kingdom and all is right with the world, the Mull was very much like a paradise on earth.

One of my favourite deliverance's to the visitors was that we could see five Kingdoms from the Mull and were the guiding Light to four of them. I left them to ponder for a while over what the kingdoms were; they would all get Scotland, Ireland and the Isle of Man; a few even got England but hardly any got the Kingdom of Heaven. That could have been the start of many a debate, so I always justified it by adding. "Just open your heart as well as your eyes, then all will be revealed". That would be as close as dare get to preaching; however, I always hoped that my tours enlightened the visitors in more ways than one.

On sunny days when there was little or no wind, I would take visitors out onto the balcony. Even on calm days the elevation would catch the occasional gust that wafted up the cliffs accelerating its force and elevating the ambiance to a level where most found it exhilarating. On the leeward side of the tower, I could show them the effect of the ebb tide as it raced out of the Bay to meet the current of the Irish Sea. The whirlpools and eddy's were easier to see on calmer days because they were not hidden by white horses dancing across the surface. The whirlpools and eddies would not trouble a boat with a good engine but could prove disastrous to canoeists. It was not often we saw them brave the Mull but when we did, they were the closest vessels to come near the Mull and I can't recollect any passing on an ebbing tide.

The number of visitors we attracted like most places was dependant on the weather; the car park was virtually empty during the winter and when it was stormy or wet. The balcony was not a place for Keepers let alone visitors on wet and windy days. Only once in my service do I remember having to brave the wind to do maintenance out on the balcony. That time I was out at InchKeith, Bob and I had to remove an old sheet that we assumed had blown across from Burntisland. The wind must have caught it and it sailed across the Forth, stopping when it hit the tower; trouble was it obscured a 6' x 4' section of lantern so it had to be removed. The wind was only blowing at near gale force, about 27 knots; so it was probably caught by a gust and carried to a higher level where the wind was a bit stronger. On a lighter note, Bob says to me in a way reminiscent of my father, "How do you get your own back?" Well I knew there had to be a punch line but humoured him just the same, "No, go on tell me." "Pish against the wind." If that was his way of telling me to get on with the task as the wind and rain was having the same effect on his bladder as it did mine, it did the trick just nicely. We had a good laugh and I still use it in my repartee on social joke telling occasions.

We had problems in the house whenever the wind became contrary, usually the wind came from a westerly direction but when it came from the north east it would swirl across the top of the roof and play havoc with the "H" can on the chimney. The smoke would retreat back down filling the living room and choking anyone sitting there. Sometimes the can would blow off the chimney altogether with the nett result being the same no matter the direction.

It was much to our relief when the Central heating engineer came to renew the system and it would be changed to gas…No more coal fires and no more blow backs…much as we liked a nice open fire with its welcoming glow, I hated having to sweep the chimney and remove a dead bird or two that had sought to seek its warmth. I ended up like a Black and White Minstrel removing a dead seagull that was stuck ⅓ of the way down, it could not be pushed down the chimney so had to be pulled up.

It was easier to clean the soot off of me than it would have been the carpet in the living room; so I did not mind the bit of cajoling she gave me when I got down.

All of that was thankfully coming to an end.

Willie, the Plumber come heating contractor and his wife come Plumber's mate Anne; were a great team and it seemed no time at all until the job was finished and surprisingly enough with little fuss and even less mess. The problem came when Willie tried to light the boiler. I had helped Willie as much as I could so it was not like I was looking over his shoulder or anything; nevertheless I could see the look of embarrassment on his face when the boiler was adamant it was not going to play ball. Anne was busy tidying up and we could see her passing the window once or twice with tools or materials in her hands. Willie was at the point where he had tried almost everything he could think of; the last course of action would be to ask Anne if she had any ideas for fear of her finding the solution and demanding he put her name first on the side of the van. Just before giving up and calling the boiler manufacturer, he thought that maybe the gas tank had been filled with an inert gas to test it prior to shipment and a residue had settled in the bottom. If that was the problem he let a little gas escape from the tank and drained the pipeline and the boiler.

The boiler decided this time it was going to ignite but that was fractions of a second after a knee-high cloud if invisible gas ignited before it. It was not pockets of gas that had collected but trouser legs full. I being five foot eight was luckier than Willie who was a good foot shorter and it was knee high on me; even still we both had the added problem of having hairy legs. On ignition a bright blue flame shot across the floor, just like in slow motion we both knew what was going to happen next and jumped as high as we could in the vain hope of escaping the inevitable flame. Just at that moment Anne passed by the window to see what she thought was our leap of joy and hears our cry of relief. She saved Willie the embarrassment of coming in to see what the fuss was about. I wondered how long Willie could keep up the act of celibacy before Anne found out the truth. All Willie could say to me was, "Don't tell Anne what happened". I politely told him there was no way I was going to sleep in another room just to hide the possibility that Mags thought I was about to come out of the closet by shaving my legs.

Willie and Anne had an uneventful time when it came to the other two houses and I learned a lesson and left all other contractors to their get on with their work and leave them alone. At least those were the orders Margaret gave me; but you know what they say about what the eye does not see.

©Steve Hardy, South Rhinns Community Development Trust

Chapter XX
High Days and Holidays

Margaret and I had been married for nearly ten years; in that time we had never had what you might consider a holiday. It never bothered us and it was something we didn't give much thought to. The girls were happy enough spending their summer holidays away with relatives. Karen stayed with Jessie and Kirsty with Margaret's aunt Maggie and Uncle Joe. Both of them would be spoilt rotten and didn't mind being separated from us, we missed them of course but it was a special treat that we could not deny them. Gavin stayed with us, so in a way we owed it to him as well as ourselves to do something special; we had Kelly to think about too, so we had to choose a type of holiday where Kelly could come.

There was little in the way of options and it did not take much thought to come up with a camping holiday. Kelly was a good traveller and loved the car and above all she was a well-behaved dog with a good sense of what was right and what was wrong.
We never gave her any formal training, when she was a puppy and up until the time she was about eight months old we tried to train her to walk to the heel and simple things like fetch and go; at times we thought she would never get the hang of it and then almost overnight everything seemed to click. It was as if she could sense what was required of her and a lot of the time she would act long before you had to ask.

Gavin was two and a bit and a lot more of a handful, it's not that he was wilful and challenged our authority, on the contrary you only ever had to say something once and the lesson was learned. I think it was more to do with his active imagination and curiosity about things that would lead him into situations.

By this time we had sold the golf to Shank's daughter and bought a Datsun Cherry. Gavin had this thing about naming things even if he could not quite pronounce the word properly, on sunny afternoons we would often go down to the beach at East Tarbert, Gavin was fascinated by the cattle that roamed free and especially the one that was different. Kirsty and Karen had called him Taurus, Kirsty was familiar with her birth sign and insisted on that name but Gavin either could not say it or deliberately chose to call him Doris just to get his own way no matter how much Mags and I tried to reason with him that Doris was not a suitable name for a Bull and if he heard Gavin, it would be like a red rag to him. Of course we had to explain what we meant, but that was making a rod for our own backs because on any future sojourns he insisted on wearing his now favourite red T-shirt. The Bull had done his job and was put to stud elsewhere so in the end he never got to see if it was true much to the relief of the rest of us. Gavin could say cherry so the name stuck, even if the car was a deep metallic blue. It was a nice wee car that had the pretence of looking like an estate car but in truth was just an elaborate hatchback, at 950 cc it lacked any real power and that was its main drawback; if you wanted to overtake anything you would have to write a letter to the person in front to stand any chance of doing so with any degree of safety. The next car we bought was a brown 1500 cc Austin Allegro estate, this was much better, more powerful yet still economical and it was reliable. This time it was Karen who chose the name Syrup of Figs simply because it was brown and gave you a good run for your money. Now I wonder whom she got her sense of humour from. The next car we got was a red Austin Princess, it was unanimous to call her Princess and she suited her name aptly, she was not the spoilt brat type of Princess but a delight to drive and comfortable to boot. The next car we bought was a red Ford Cortina Mk5 estate, not only was it the newest car we had bought but it was to last us until I left the service.

Gavin got to choose the name for this one and because it was red and reminded him of the T-shirt and the Bull, he called it Doris.

We bought a tent and all the paraphernalia that goes with it and loaded it all into Cherry. We chose to spend our holiday camping near Aviemore and went via Killin to stop overnight at a campsite by Loch Tay. The scenery as expected was beautiful but the water of the Loch too cold to go dipping your toes into. I think the name Tay should really be Tea because that was what the colour of the water was like and gave the Loch a mysterious darkness to it; blend with it the fragrance of peat and pine and you would definitely have a good Earl Grey. I'm not sure how deep the Loch is but they say that from the Loch to the sea the river Tay has the greatest volume of water passing through it of any UK river. I used to wonder if the Loch would be deep enough to both shelter and hide a Loch Tay monster and the locals may have missed their chance of fame in the past. Upriver from the Loch is Aberfeldy with its notable distillery. To the connoisseur each whisky will have a distinctive flavour that is created by the source of its water and it's colour derived from the casks in which it is matured, so the colour of the water from the Loch bears no reflection on the final colour of the whisky but I think some Whisky Buffs might describe Aberfeldy malt as being the Earl Grey of whiskies.

We joined the A9 at Pitlochry and stayed on that road till we got to Aviemore, both Margaret and I had seen much of the scenery before on our train journey up to Holburn Head; the railway line follows the road for much of the way. Once we were at Aviemore we could use it as a base to visit other places like Inverness, Loch Ness, Nairn and Fort William and Local sights like Boat of Garten and Granton on Spey.

If I were a keen fisherman and I could afford the permit then The River Spey offers the finest salmon and trout fishing in the country as I was handicapped by one it made me less interested in the other and I much preferred to take in the marvellous scenery. Apart from Mags this was what I had moved to Scotland for the chance to see. The Southern Uplands has its own majestic beauty to be sure, but is dwarfed by comparison to the Highlands. The Alps and worse still the Himalayas are barren deserts but raised to lofty heights where only the tracks of man are foolish enough to tread its peaks just for the sake of it. The natural beauty of the Highland ranges has been left to get on with it, much of its wilderness unseen and barely touched by man.

I am letting my imagination run away with me because I have not seen what lie over the next hill, in the next valley or over the next rise; all I saw was what was before me and I was fulfilled.

Gavin got up one early one morning and managed to find his way out of the tent; we had secured the tent zip with a lock to stop him from doing just that, but young Harry Houdini had got out somehow. We got woken up by the sound of banging coming from somewhere nearby, we thought it strange for someone to be working at six in the morning and went out to investigate. We were not the only ones to be woken up but we were the ones horrified to see our Gavin banging a stick on the side of a mobile home. I had to dress as quick as I could and rush down to stop his antics and apologise profusely to the owner. Luckily he saw the amusement in it and was able to forgive the little mischief but only if he did not do it again, there were other complaints from other site users that took a little more to defuse. I had wondered why Kelly had not bothered to tell us that Gavin was outside the tent, perhaps even she was as bemused and horrified as we were but chose to distance herself by having that "He's not with me kind of expression on her face". We got to see all that we wanted to see, Gavin would have been as happy sitting on the front step playing with the gravel but we think he got as much out the holiday as we did.

Our next holiday was a year or two later and this time we took the girls with us. We had always fancied the idea of owning a caravan and had the notion of travelling around the UK seeing some of the places we'd always wanted to see. We both felt it prudent not to go spending too much on a caravan to start with and to choose a place that was not too far away; after all I had never towed a caravan before. First thing we had to do was buy a second hand caravan; there were quite a few caravans for sale but the only one we found suitable for our needs at the price we wanted to pay was at a caravan site in Cairn Ryan. At the time we only had Syrup of Figs and I doubted she could tow the one-ton Linton 6 Berth caravan the 32 miles to the Mull. The least I could do was try, the terrain was pretty flat until you got to Ardwell but the real problems started just outside Drummore and climaxed in the 1 in 10 hill going up to the Lighthouse. You could hear her struggle if you had to slow down for some reason before climbing a hill but she was fine if you could get up a head of steam on the approach. One thing was for sure and that was there was no way that she could handle any great distance even if it was flat all the way.

When we got to Tarbert without old Syrupy dying from the strain; I really thought there was a good chance we would get to the top without stopping. That would have been the case had it not been for a stubborn old cow that chose to walk the road just out of sight on the bend half way up the hill. I had to break sharply and had visions of the caravan becoming unhitched to find its own way down the hill again, at best careering of the road to an inaccessible place where I could not hitch up again or at worst with ever mounting velocity career off at a tangent and go sailing down the embankment to be dashed on the rocks below. Thank god this is not the script for a Carry On film otherwise that's what would have happened. I only had Mags with me at the time, so I let loose a few choice profanities at this belligerent bovine; she had now taken to standing stock still in the middle of the road. My biggest worries were whether old Syrupy could tow the caravan from a standing start and if the cow remained there much longer, I risked encountering another car coming down the hill. Not an ideal situation to be in without any passing places and the ground uneven either side of the single track.

There was no way Mags was going to get out of the car to shoo the beast away and there was no chance of me getting out of the car just in case the handbrake failed; So it was a Mexican Standoff and we just had to wait. I was thinking of a hundred ways to eat beef and lost count in the late twenties before she decided to move out of the way.

Good old Syrupy, even from a standing start she managed to pull the caravan up the hill. Syrupy had given us an extremely good run for our money and we even got nearly as much as we paid for her but alas she lacked the weight and the power for the task she had to perform.

Her replacement came along in the shape of Princess, a red 2 litre Austin Princess with a black soft top. For her age and the price we paid for her, she was a comfortable driving experience. Her imitation leather seats had drawbacks unless treated to some suitable covers. Damned hot in summer and freezing in winter; the good thing about the driver's seat was that it was adjustable to my ideal driving position. I bought the car on the understanding that the garage fit a tow bar.

I had done quite a bit of reading on the subject of caravans and how to handle them on the road and a tow bar was just one of the essentials needed for towing. I also thought it safer to fit a stabilizer unit; after all I was carrying the most precious cargo of all I could afford to lose the caravan but not the family. There was no sense in me buying a heavier car to match the caravan if the end resulted in the caravan forcing the car off the road if it caught a gust.

I had to buy extensions for the wing mirrors, so all in all the price of our caravan holiday was mounting by the week and we had not even decided where to go yet.

We knew that if we gave the kids the choice of anywhere then they might struggle so we narrowed it down to either the seaside or the countryside. We cheated a bit there because we knew the kids would choose the seaside, Gavin almost threw a spanner in the works by first of all saying countryside. He was developing a sense of humour at this tender age of four and used us as his guinea pigs; he had picked up an expression from the girls and used it as his punch line, "just kidding." The Royal We was not amused. "That's settled then, we're going to Morecambe" I said; Mags stared daggers at me and I could feel her elbow digging into my ribs and the muttering under her breath, "Since when".

Sometimes you have to peel back the layers before Mags can see what lies beneath, normally we would discuss everything in full before a decision was made but in this case I had to make a judgment based on what was best. I'd chosen Morecambe because of its distance and that it was before the more expensive neighbour Blackpool. The site I had chosen was within our budget and it had a good range of amenities. Ocean Edge was in Heysham not Morecambe but close enough not to make much difference. The holiday park catered mainly for static caravans but there were designated pitches for caravans; some of them even had power points.

I had taken the time and trouble to wire the caravan up for electricity to supplement the battery; however, naively I had not considered how the supply would be brought to the caravan. There would be an additional cost for the link up cables as well as an exorbitant charge for the electricity itself. In the end I chose to park the caravan on a pitch just under a street lamp, the light was diffused by the skylight so it was not too bright and shed enough light for the children to see at night; what's more it was free.

They say a change is as good as a rest and for Mags and I that's what a holiday was all about. We still had the daily routines to perform; only the location was different with a modicum of the novel thrown in for good measure. We all got something out of the holiday even if it was not all good.

The children enjoyed themselves at the Western World theme park, Gavin and I loved the trains at Carnforth railway museum, Mags was a bit disappointed when we could not find Emerdale; I was disappointed when we stopped in Harrogate for something to eat but found all of the fish and chip shops closed for lunch…"Strange People" I thought…What spoiled a lovely picnic for us at Middleton Sands was the interruption by a pair of half crazed Irish Setters who brought another meaning to the word sandwich or should I say sand wedge. What God gave in looks he took away in intelligence and I'm not just talking about the dogs. After a poor excuse for an apology their owner made things worse with her parting comment of "Enjoy your picnic," before running off after them. I wish we had brought Kelly with us she would have given them what for; she was in no way aggressive but she did not tolerate buffoonery and knew just how to put Gavin in his place. A well-timed growl with a menacing look of intent was enough; of course it was all a show of bravado because when push came to shove she would hide under the covers rather than face the danger. Some guard dog she would make, there was more than one occasion when she heard something that might pose a threat and she would give a muffled bark, just loud enough for us to hear and hopefully wake up but not loud enough for the possible intruder to hear. If that didn't work she would bark just a little louder, if she still heard a disturbance after that and we had not woken up, she would dash under the covers just in case it was her they were after. I still think she would have made a stand against these two interlopers and been able to share some of the picnic with us.

Gavin's favourite place was down at the bouncy castle, he was there so often that he was on first a name terms with the manager, we were sure that he stayed on longer than the hour that he paid for. Mags had to go and get him one evening for his tea, in a bit of a hurry she picked up his trainers in one hand and grasped Gavin's hand in the other. After tea we were going for a walk, so she went to put his trainers on; his left shoe fitted fine but when it came to the right shoe no matter how hard she tried the shoe just would not go on, frustrated she says to Gavin, "what kind of feet have you got?" Seconds later it dawned on her that one shoe was slightly smaller than the other. She had picked up some other boys trainer by mistake; so there was probably another mother going through the same rigmarole as she was. Sheepishly she took Gavin back to the bouncy castle in the hope of finding the other shoe still there, with a sigh of relief the owner of the shoe was still playing happily, so she quietly replaced the shoe and lifted Gavin's, she tied it so tightly that Gavin winced; slackening the knot she says to Gavin "That's it, there will be no more bouncy castles for you young lad." He was back again the next day and every day after until we left

The holiday was a bit of an eye opener for us and played a good part in moulding our ideas for holidays. A lot of the Static caravans were brand-new and all had the fittings and comforts of home, worst of all, when compared to what we had paid for our two weeks, they had the benefit of a lot more luxury for a relatively small amount more.

The future for us was not in Touring neither was it in Holiday Parks we unanimously agreed that we should stick to our paradise home at the Mull and the girls could go back to being individually spoilt by Jessie, and Maggie and Joe.

Our holiday had to be cut short because Kirsty had come off a slide at the Park and split her bottom lip wide open at the chin, apart from the stitches she needed emergency dental treatment and that was best provided by her own dentist. Maybe that was the final straw, the decision made unanimous every time we looked at Kirsty's face.

The Long and Windy Road to the foghorn

Chapter XXI
CB Radio and The Russian Ships

Ron introduced me to the world of Citizen Band Radio; Mags would not forgive him for it, because I became hooked the minute I heard the voices chatting away over the ether. To start with I would listen in on Ron's CB, he kept his in the workshop and had his antennae mounted on the chimney next to our TV aerial. We were in an ideal location for transmitting and receiving and were able to talk to folk far beyond the twenty miles that the wave band was designed for. In the folklore of CB fanatics it was said that it was possible to send a message round the world on a ¼ watt of power. I'm not going to dispute that, but I don't think there is anyone alive who has seen or heard of it. In theory it is possible because every so often the signal can in character change from VHF Line of Sight ground waves to HF Sky waves that can bounce off the ionosphere and so travel greater distances. In CB terms it is called skip and it was every CBer's dream to speak to someone on the other side of the world. True enthusiasts would send each other cards to acknowledge the transmission; these were called QSL cards. I have forgotten a lot of the terminology of the airwaves but just like the marine band there was a frequency or channel for calling somebody and that was channel 14. If you wanted to call somebody up you would say, "14 for a copy."

Normally if you were new to CB it might take a while for someone to answer your call simply because you were new and might be considered to be a plonker. I remained a plonker for a few weeks before I realised it was better to try and introduce yourself by breaking into another conversation.

In marine band terminology you would have a call sign, in CB terms it was your handle. My first handle was Pulsar in reference to the lighthouse but in the end nobody seemed to get the connection so I took the drastic decision to make my handle more memorable, so chose the Galloway Flasher. There will probably be some CB enthusiasts who might think I was still a plonker but in the main the name change attracted a lot of attention and I began to talk to the same people on a regular basis.

I could not keep using Ron's radio so I bought my own rig; I had taken some advice on buying a few accessories such as a SWR meter and a boom microphone. The SWR meter could check the output of your signal, if you kept it within tolerance levels you would maximise the efficiency of the signal. The boom mic or microphone amplifier could, if you used it properly enhance your voice to make it clearer; on the other hand if not used properly it might have the opposite effect or make you sound as if you've got your head in a bucket. There were a couple of times that I had inadvertently touched the control and had someone come back with that expression.

I liked to have a good old chinwag or just listen in but found as much entertainment in both. If you were trying to join in a conversation you would at a convenient moment say, "breaker on the side," if they knew you or wanted to know you then they would reply, "breaker on the side come back." The term breaker came from America I believe and made reference to the fact that the CB radio was illegal there at one time so anyone using it was breaking the law, in the early days the same might have applied in the UK so the name stuck as did a lot of the other words.

I was fortunate to make radio friends with people from all the compass points and just like the range of the light, be heard and talk to folk from over 60 miles away, the greatest distance I recorded without Skip conditions was to a chap who was on Snowdon in Wales and that is just short of 200 miles away from the Mull. I would talk sometimes with another guy at Southport and another from Banbridge in Northern Ireland. More often than not I would be kept company during my watch talking to my regulars, the Scottish clan; Red Eagle, Daffodil & Lobster, Daffodil & Rosebud, White Ghost & White Satin, See thru nightie, Blue Lagoon, Tractor Driver and Tartan Lad. The English contingent of Mary Poppins & Tango 41 and Black Tulip; and from the Isle of Man; Green Fingers, Shaun the Hannibal and Pole Cat. There were a few from Northern Ireland but the only one's handle I can remember was The Mechanic and if I remember correctly he was engineer on board a coal ship. The Black Tulip was master of a coastal tanker, he was a Dutchman and up until the time that the British Seaman's Federation had the British Seamen for British ships enforced; I could regular as clockwork speak to him from the moment he left Silloth in Cumbria all the way until rounded Corsewall Point into the Firth of Clyde.

We used to have a great laugh and there was always a busy channel when Jon was on the airwaves; when the weather was right, he would pass as close to the Mull as he dare and the whole family waved to him as he sailed by; all we had to do was step outside the back door. At that time I was broadcasting from the kitchen so all I had to do was lift the rig and microphone and I could talk to Jon and wave to him at the same time.

There was one drawback of using the kitchen; nothing and I mean nothing was going to stop Mags from doing the washing; so when it came to the spin cycle I used to tell my listeners that I was, "Going mobile on the washing machine," usually going mobile meant that you were transferring your rig or transmission to a car, but in my case the whole rig used to move with the vibration and it became impossible to talk or listen. Mags was close to becoming a CB widow until I began to see sense. In the end I restricted the CB to when I was on watch and the family were safely tucked up in bed.

Mags and I went to a CB rally at a caravan site near Kirkcudbright called Auchenleary, we had the company of Mary Poppins and Tango 41 who were visiting the area and we invited them to stay in our caravan. Apart from meeting other enthusiasts from all over the region and afar, there was a general knowledge quiz for teams of four. We called ourselves the Galloway Flashers because unbeknown to me my reputation had spread far beyond our corner of the Irish Sea and a more people than I ever spoke to had heard of the name Galloway Flasher.

I had always been a bit of receptacle for useless bits of information but this was the first time since my school days that I had been in a competition; however I had team members and a reputation to think of, so I hoped that I would put all that stored up knowledge to good use. It was close and in the end it came down to a tiebreaker to which the answer was German Democratic Republic. The prize was not much but it meant a lot to me then as it was the first time I had ever won anything. Where the small trophy is now God only knows.

We used to get a CB buff and his wife come down to the Mull every year, they came from Haltwhistle in Northumberland, I can tell what you're thinking but even with the best rig in the world I would not be able to speak to Harry and his misses from here at the Mull and it would be a one a million shot for Skip to hit that particular spot.

That was the peculiarity with Skip, the area it affected was limited and the best way to describe it was that it was like a rainbow, the only people that could hear or speak were those at either end and just like a rainbow it was there one minute and gone the next. There was one time when Harry had made use of our kind offer of the caravan instead of the cramped VW dormobile that he had. The weather was hot and sunny and the Sun was going through its eleven-year sunspot activity cycle so the Skip was coming more often. The other end of our rainbow was the south east of England from the south of Essex to the channel coast of Kent. There are a lot more breakers in that part of the country and it was hard for me to break in on the side; for one I had not lost that much of my accent so it would have been hard for them to believe that I was talking to them from Scotland and two when I was fortunate enough to talk to them it was only briefly because the Skip had moved like a wave further round the Essex coast and out into the North Sea.

What was most remarkable was that Harry's rig was no more than 50 feet away from my own but yet we could not talk to each other; but what made it most memorable was that I was able to speak to an old friend. Jon the Black Tulip had to take the lesser position of third officer on a Norwegian wood cargo ship and their main route was between Norway and France. I recognised his dulcet tones even when he was speaking his native tongue; I could not begin to believe that he would be able to hear me above the clamour of the rest of the calls of QSK and CQDM…the more formal request for breaker on the side. If you want to know what they stand for then forgive me but you will have to ask a true enthusiast which was more than I was; all I was interested in was being able to talk to someone on the other end of the radio. I could hear Rosebud repeatedly calling up but without any luck and I feared that I would do no better but Jon answered on my first call. Many guys like Rosebud would have thought it a coup but I enjoyed it for what it really was, me talking to an old friend. I was glad that the Skip lasted long enough for me to get all the latest news from Jon and I am sorry for all of my fellow breakers who tried in vain to speak to him.

The closest I got to be a buff was when I started making my own QSL cards to send to people. I used felt tip pens and drew a figure wearing a kilt with a lighthouse in the background; the kilt lifted up and underneath it was the wick of a candle; apart from the greetings the motto said if you want to know what a Scotsman wears under his kilt then lift here. On the underside of the kilt it said you'd get your fingers burnt.

It all sounds a bit ketch now and perhaps I should be a bit embarrassed to be associated with them, the only way to know whether they had any significance to the people who received them would be for them to be recognised and turn up again.

They were happy times and I hope the Galloway Flasher is remembered for all the good reasons and I hope that I bought a little happiness into the lives of my fellow breakers. Sadly this old breaker is broken and put away his radio for the last time back in 1989 when he was sent to a place where the breakers were few and far between.

Today I live in a valley and would have to wait for the Skip to come for my signal to escape, I neither have the money or the inclination to go into the more adventurous Amateur Radio side of communication so the Galloway Flasher must be consigned to the past unlike the Light on the Mull that will shine on until the need for it is consigned in the same way.

Ron also had an interest in keeping his eyes open for Russian ships. We were still in the grips of the cold war even if it looked like global warming was causing a bit of a thaw in relationships. Gorbachev and Reagan had a kind of relationship that brought hope to the world. When Gorbachev came to power in the USSR it was said that the birthmark on his forehead was a tattoo of the USSR, later we got to know that it was his beloved Georgia that was ingrained into his heart as well as his mind.

The interest for the Russians in our neck of the woods was the Submarine base at Faslane and the Firth of Clyde. There was little to stop the Russians from seemingly using cargo vessels with a legitimate reason for using the Irish Sea and the North Channel as a passage into the Atlantic while at the same time monitoring the activity of our nuclear submarines; that is supposing they had the equipment on board to do so…nod nod wink wink.

During the heart of the cold war I was serving in Germany; I can remember being on exercise or manoeuvres, I'd prefer to call it exercise simply because once we were in location we did not do much in the way of manoeuvring; and we were close to the East German border. Word came down that there were SOXMIS in the area…Soviet Military Mission…Legitimate spies for want of a better word; as long as they made their presence and intentions known then they could observe. I suppose we had the same people doing the same thing over there. If we did spot any we were to report it as soon as possible so that we…British Military Mission…could keep an eye on their spies. Makes you wonder does it not that if there were enough of them keeping an eye on each other then we could have all gone home. Forgive me if I sound a bit less than patriotic but you have to remember the regiment I was in was a bit of a white elephant. I would have gladly returned to the colours during the Falklands crisis had my services been required. Anyway getting back to the exercise; what I tell you now could not be admissible as evidence as it was strictly hear say but as the person relating it to me was not given to exaggeration and as he was a lot bigger than me I would in no way call him a liar, just imagining the sequence of events still has me in fits. We were coming into action after moving locations and John Hall was on guard duty; the weather was foul and it rained throughout; to make dress matters worse the Brigadier had decided to make us come into action wearing our Nuclear Biological and Chemical Warfare Suits as well as our Gas Masks. It was at times like these that I was glad I became a radio operator, it only took me twenty minutes at most to come into action; although I could take my Gas Mask off I had to keep my suit on, but at least I was in the warm manning my radio. I was the only person allowed not to wear a gas mask; as yet no body had invented the throat microphone or it was too expensive to dish out to the operators. Thinking on it now, had it been the real thing I would have died manning the radio while the rest of the Troop survived because they wore their gas masks. We tried communicating with our gas masks and our NBC suits once but whatever muffled messages we got were alien and unintelligible like…alewy twarf oneefe aphhawah… fart (exhale of breath)…chisee estw choo …fart (another exhale)…wardioi cherked offay …fart. Translates if you heard it in the first place as Hello two one alpha, this is two, radio check over. Sorry about all the farting but that was the sound of escaping breath through the side of the gas mask. The Battery Captain either had an acute sense of smell a wild imagination or the good sense to call off the exercise as a complete waste time.

Now I bet you're thinking that I am a bit like Ronnie Corbett when it comes to telling a story, OK, OK I'll get to the point. Now there's John Hall on guard with Gas Mask on and his NBC suit and on top of that he is wearing a waterproof poncho. Even in the Army they had oversized ponchos to dwarf even John's large frame, it was not that John was fat, what John had was all muscle and he was lean with it; his poncho concealed his SLR…Self Loading Rifle…that he had slung un-soldier-like barrel downwards from his shoulder. He would have got a quick reprimand from the Troop Sergeant Major and even worse from the Troop Commander had they caught him.

While he was on guard he noticed this car pull up and turn off its lights, John immediately thought it was SOXMIS so rather than report it straight away he thought he would scare the shit out of them. He crept up to the side of the car as quietly as he could and peered through the back window where he had heard voices. It was not SOXMIS that he had scared the shit out of, but an innocent yet not so innocent couple in the throes of lovemaking. Within the space of a few screams of terror the man had leapt back over his seat in a feat worthy of an athlete, leaving his mate in the back in whatever mess they had created between them and drove off at high speed; the figure of John Hall looking even more alien like in his rear view mirror as he tried desperately to un-sling his rifle.

We had a pair of binoculars and an old telescope at the Station; the binoculars were OK for seeing ships and gave you a rough idea but when it came to recognition of the type and origin then the telescope was the tool for the job. Ron reckoned the more Russian ships that were recognised and confirmed the better chance there was that we could get a better telescope. I was all in favour of helping him achieve that goal; now that there were two of us at the Station with an interest we could get an updated version of the recognition book.

There was a regular Russian visitor up the Irish Sea and that was in the shape of a RO-RO car trans-porter…Roll on Roll off…At that time all Soviet Ships had White funnels with a red band; once you saw the red band you could then look at the recognition manual a to find a match in the superstructure. You did not have to see the Hammer and Sickle emblem within the red band to know it was Soviet. They were not the only ships we saw but we were lucky to spot one or two a month. It's not surprising when you come to think about it, if Russian intelligence was half as good as it was made out in films then they would know when the nuclear subs left their base or even perhaps when they were due to leave their base and so could have one of their vessels on hand to monitor. This is where I get more than a little cynical but I bet they knew we were watching them and we knew they knew we were watching them, just as long as no one has the will to press the trigger and we are still in a position to be able to afford it; then they can play their games; that's my way of thinking.

Ron got his wish and a new telescope arrived before the Reykjavik talks began and the prospect of ending the cold war became a probability rather than a possibility.

©Steve Hardy, South Rhinns Community Development Trust
A picture of Ron Ireland and his family now on display in the museum

Chapter XXII
A True Family Station & The Price of Progress

It was a sad day when Ken retired from the Mull; most of his furniture was already at Hunters Moon, so there was little to do except say goodbye. We would miss him and Mrs Clark but we would make a point of visiting them on our way to Stranraer and later on when they moved to Glenruther Lodge at Bargrennan. Neither was out of our way but the Glenruther Lodge route offered the best in the way of scenery as the route took you passed Loch Trool and Glen Trool following the Southern Upland way for about 15 miles before veering off over the hills to Straiton and onwards to Dalmellington and Patna.

John Lamont arrived as Ken's replacement and he brought with him his wife Agnes and daughter Alison, more importantly Alison was not much older than our Karen, so not only would they have a new playmate; but there would be someone else to share the trip to Tarbert for the school run. John came from a Lighthouse family and both his brothers were Principal Keepers; Murdoch was at Neist Point and Hector was due to go to Corsewall Point at the other end of the Rhinns. Hector later went to the Mull of Kintyre and was Principal at the time when the Chinook helicopter crashed killing all on board.

I never got to meet Murdoch but if he was anything like his brothers; I think it would have been a treat to see them all together. John was the oldest of the three and was the more serious or so I was led to believe from personal experience and from the exploits told to me by Hector, who made frequent visits to the Mull. Now there was one man I would have loved to have served with, we both had the same kind of humour and he was most likeable in his bearing. That's not to say that John was any less likable because John was a damn fine Principal but he took himself just a little too seriously. I'll give you an example; they had a spaniel called Struan, a lovely tempered dog but not quite the full shilling in the intelligence stakes; I used to throw a ball for Kelly that went over the wall and down the road towards the garages. Kelly would run after it but use the inside perimeter wall as a springboard to get over the higher boundary wall at the apex of the two. John says to me, "Struan can do that," so he throws the ball in the same direction, Struan thinks about it then dashes after it following Kelly's example but instead of going over at the apex, Struan goes over where the local's caravan was and went thunk straight into its side. I could not help but laugh, but John took the huff and stormed off back to his house.

Struan redeemed himself though and brought a smile to our faces when he did with his nose what Drummore Stores cat did with its tail, both John and I would look the other way disowning Struan in the process.

Agnes Lamont shone as bright as the light at the Mull, her love of life and the people around her effused from her very being. You could not but want to get involved with every venture that she undertook. She was Mrs Motivator in the kindest possible way. Although we men-folk had our work to keep us at the Station, it was not long before Mags and Anne were enrolled in the young women's group, The Mull Women's Rural and keep fit classes down in Drummore. Agnes could drive and John reluctantly let her take Poppy his pride and joy Lada out for a run. Now don't go mocking, you can get very attached to cars even if Poppy was a Lada. She may have been a cheap version of the Fiat but she proved as reliable as any car I ever had and could happily join Syrupy in my annals of motoring history.

Through Agnes and Alison the girls got involved with the Brownies and the guides so they got out and about more as well.

Compared to Agnes, Alison was more like the balance between her Mother and her Father. To me she seemed shy and remote but she was like that with all adults outside of her immediate circle.

Gavin had a great fondness of Agnes and he would often accompany her when she took Struan for a walk; he also had a liking for Thomas the tank engine, so they would play at trains; using the pathways as the branch lines. Gavin had to be Thomas of course, Agnes was Gordon and poor old Struan was Percy. Even at four years old

Gavin was able to recognise that Struan was not quite all there. As for Kelly she stayed at home it was beneath her dignity to be seen playing such games. She tolerated Struan but that was all, there was no way that he was going to get close enough for there ever to be little Kelly's and Struan's running about the Station.

The nearest Kelly ever got to losing her virginity was when I picked up a stray Jack Russell on my way to collect the girls at the end of their summer holiday. He was walking boldly down the road not a care in the world, I took an immediate liking to this chap and though it would be a shame for him to get picked up as a stray that was undoubtedly what he was. With just a little growl he allowed me to pick him up and put him in the car. We all liked him and it looked as though he liked us too; we had not only provided him with a roof over his head and a steady supply of food but a bitch ready and willing to offer her services when she came into season, what more could he ask for.

We decided that we really did not want pups about the house, so Dougie would have to get the snip. I took him to the vets and duly paid the fee, however, they could only find and remove one testicle the other had not dropped so it would need to be taken out at a later date. Dougie must have thought there was no way he was going to lose his last chance of sowing his oats; so he skedaddled the first chance he got. He still had his lead on when he escaped at the Mull. I do hope that he managed to break free and returned to his life on the road but sadly I think he may have chased after a rabbit and disappeared down a burrow getting stuck. That taught me a valuable lesson that sometimes even an act like a Good Samaritan can have consequences.

Instead of a litter of pups we had a litter of kittens; we had got Jess to do the job that Tats was good at and that was keeping mice out of the house. Gavin called her Jess because she was black and white just like Postman Pat's. She was every bit as good a mouser as Tats was, but lacked his temperament. Her litter arrived during the middle of the night and we awoke next morning to see her nursing them in the tank press.

We had hoped to see one of Tat's offspring among them but although she must have mated with more than one Tom, none had Tat's tabby markings. Two of them were almost identical and were slightly larger than the rest; we called them the Sable Twins, an almost pure black one we called sooty just because his coat was a bit scruffy, another we called blaze and she was like her mother, the runt of the litter we named Cat Weasel he may have been small but he was the one with the most mischief in him. We were lucky to have found good homes for them all; one cat in the house was more than enough.

Our problem or should I say Mags' problem with mice was that she could not stand them and squealed like a school girl every time she saw one. It came to a head when she was looking through one of the kitchen drawers and came across a box with something in it, thinking it was a sinker that I used for fishing she emptied the contents into her palm only to squeal hysterically on seeing the mouse. Just before we got Jess we had to call the electrician out to sort the cooker that continually blew fuses. He found the remains of one electrified mouse stuck in the wiring at the back of the cooker. We generally only had mice during the autumn when they came into the house looking for a bit of warmth for hibernating. After we got Jess we did not see any more mice. Jess was friendly towards Kelly but there was not the bond that there was with Tats, when the kittens were a little older Kelly would baby sit while Jess went for a wonder. We always thought Kelly would make a wonderful mother and the way she treated the kittens was the closest she got to prove it.

Mags and I got out a bit more often and we had made friends with a family from Logan Gardens. Teresa met Mags at the nursery in Drummore and they became firm friends, I was introduced to Stuart through that friendship. Stuart was about the nearest I could get to anybody to be able to call them friend more than just colleague. They enjoyed coming to see us at the Mull and we enjoyed seeing them at their home in the Gardens. They had this amazing Airedale Terrier called Tweed; he was friendly and full of character and seemed to love everyone. All of their children were a credit to their parents and showed the same affability as tweed did.

We would often meet up at some of the local functions in Drummore, every year they would hold a Halloween Party, those who chose to dressed up in fancy costumes; Mags and I just took pride in making costumes for the girls to wear. One year I converted an old sheet into a pair of outsize knickers, Karen we made into a daffodil and Kirsty we made into a daisy. With one down one leg and one down the other we had made a pair of bloomers, they got a prize but had to change into the spare clothes that we brought so that they could play with their school pals. Mags made the clothes for future Halloween parties.

Kirsty insisted that I make her a costume for a monster birthday party that she was asked to. I made her a wonderful Dragon suit out of cardboard and took a lot of time and effort into its construction. Kirsty was proud of her fancy Dragon suit but her Dad should really have thought of something a little more practical for a girl with Kirsty's tiny frame. She won a prize as well but could not enjoy herself without removing her suit, if I had thought about it she could have worn the same kind of raffia flower head to transform her from a dragon to a flower in one easy step; after all she was wearing a one-piece green leotard.

Mags went to an auction held to raise funds for the church, I could not go because I was on watch so I stayed and kept an eye on the kids. Mags came home with a picture of a naked couple seemingly held within the wings of a giant swan; the picture was called wings of love. She also came home with a metal coffee percolator we never used the percolator but the picture hung in our bedroom until Gavin noticed what he thought was the mans willy, in truth it might have been his other hand but what with that and the fact that the Minister kept asking her if she still had the picture was enough for her to have me consign the picture to file thirteen.

All of our years at the Mull were happy ones but these were the happiest of them all, to crown it all we even had snow at the Mull and it lasted long enough for the children to have snow ball fights and long enough for Ron and I to make sleds for them before it melted consigning the sleds to the next bonfire we had in the garden. For five days we had a cold spell that froze and burst the pipes of the water tank on the hill. We had to do makeshift repairs until the plumber could get out to fix it. The roads were treacherous from Drummore to Stranraer and not for the first time the older Children who went to the academy in Stranraer had to be boarded out. Karen and Alison were due to start the academy in the summer but for the moment the minibus was taking them all to the Primary in Drummore.

Kirsty was lucky to escape injury when she fell asleep in the minibus on the way home. She was leaning on the back of the front seat when Mr Shanks pushed it forwards to allow other children off. Kirsty was catapulted over the seat and out of the door in the blink of an eye. Mr Shanks thought that someone had kicked a ball off the bus and was not amused when a bedazzled Kirsty got back on. "Where did you come from?" he asked. I think Kirsty was still too dazed to answer and it was our Karen who told us the story when they got home.

We knew that our time at the Mull would have to come to an end sooner rather than later, we had already been there for just over five years. When it came it came in the most sudden and devastating fashion. For a long while behind the lighthouse scenes there were talks going on about the future of all lighthouses around the coast of UK. With the advent of new technology there were arguments whether the service we provided offered the best value for money. Justification for funding was becoming harder and harder until it was decided that a reduction in manpower would be the best way forward in order to reduce the revenue needed to keep the service going. One-way of achieving that would be to discontinue the fog signal at most of the Stations.

In effect that would reduce the Service manpower by almost a third and so consequently reduce the wages bill. For the likes of the Mull it meant that because of the nature and source of the light it would take very little cost and effort to automate it; in the meantime, the Station without the foghorn could be manned by one person. The Local Assistant and one of the Occasional Keepers would be made redundant, Ron and I would be transferred to other Stations, leaving John to stay here until the automation was complete and he could take retirement

It did not take much working out that my days as a Lighthouse Keeper let alone my days at the Mull were going to be numbered; it was no wonder that the tragic news brought tears to my eyes. It was to take immediate effect and so that day was my last working day as a Keeper at the Mull.

It was only a short time ago that Ron and I had cause to give aid to a boat in distress; it was at change of watch at ten o'clock. I was coming on watch and did my usual thing of looking over the wall out to sea just to get some idea of what the weather was going to be like. I could see a bank of fog drifting closer to the Station but not yet at a distance where the horn needed to be sounded. I could also see a boat drifting closer to the Station than I thought was healthy so I started the engines to sound the horn as warning. Ron had heard the engines start and could see me opening up the valves to the horn; Ron could immediately see the reasoning behind it and completely concurred without need for him to remind me it was still his watch. Just after the first blast the people on board set off a distress flare; now we knew the reason why they were drifting toward the Mull. Ron telephoned the Coast Guard and I kept watch on the vessel just in case she got into more difficulties. I was amazed when Ron came back to tell me that the Coast Guard were aware that a boat was having difficulty just off a Lighthouse but they did not know which one. Well they knew now but what beggars belief is why didn't the boat carry a marine chart and why didn't the Coast Guard tell him to count the flashes of the light.

They sent out the lifeboat from Portpatrick to tow it to safety and an auxiliary Coast guard to the Mull to make a report. The young lad they sent out was as bad as the guy on the phone and seemed to know nothing of the geography of the area; we had to inform him about all the visible landmarks right down to the basics of where north and south were. Had the vessel been a cruise liner or an oil tanker then they might have had second thoughts when it came down to discontinuing the fog signals; the trouble was that the people who needed our service the most never paid a penny for it and it would be unworkable in practice trying to collect money from them as well as a change in the act of Parliament.

We often had a visit from the local customs and excise officer and he would ask us if we had noticed any unusual activity at East or West Tarbert either location would be ideal for landing contraband as they afforded good shelter and were obscured from all points except the Mull. Most of all there had been Lighthouse Keepers at the Mull since 1828 and they were valued amongst the community of Drummore, soon there would be no keepers at all; and although the light would still shine the heart and soul of the service would be gone. Maybe in fifty years time the Mull will still be shining for posterity but she would be a heartless light shining out to warn a crewless ship.

A year ago the Northern Lighthouse Board celebrated it's Bicentenary, we even had our photographs taken for the book At Scotland's Edge that was published for the celebration, now it felt as though we the Keepers who had done most to make the service what it was were about to be consigned to the archives of history books and the documentation decaying in the vaults of 84 George Street. Yes there would be a few years left for me, another 5½ to be precise but they would be my last in a service that I loved and they would be the hardest to bear.

Ron was going to Killantringan for a while and then he would be going up to Oban as a Relieving Keeper; I on the other hand was going to Ailsa Craig.

We had only seen Sandy Wright once before we got the news and he was a newly appointed family liaison officer; perhaps that was one indication of troubles that lay ahead; he was one of us so it was easy to confide in him even if he was as much kept in the dark as the rest of us. Now that the cat was out of the bag he had his work cut out and was making regular trips to lighthouses most affected by the cuts. Wherever possible the Board would relocate Keepers to Lighthouses of their choice, it was an easy choice for me to make when it came to Ailsa Craig, the shore Station at Girvan was only seventeen miles from Patna.

The plan was for me to go to Ailsa Craig as soon as the Assistant there had moved; he had taken redundancy and had bought a house but there was trouble when the deal could not go through. So I had an extended fully paid holiday at the Mull until it was sorted out. There must have been a lot of problems around the service at that time involving the same kind of expenditure but once it was sorted the savings on a reduced workforce would more than balance things out.

As it was, Ron and Anne were the first to leave the Mull it was sad to say goodbye especially under the circumstances. But we wished them well, Ron would probably get enough years to see him attain the same pension rights as if he made Principal, I had chosen to take voluntary redundancy when it came my turn in the order of seniority and at a time that would offer me the best reward in terms of settlement.

We still had the caravan to think of and the Doris had replaced Princess as our car but she didn't have a tow bar.

Princess had an argument with a Vauxhall Corsa driven by a young lad who had not long passed his test. The road we were driving on had just been resurfaced and there were a lot of chips loose on the top, the lad came hurtling down the road at a speed well in excess of what he could safely handle given the conditions. The worst thing he could have done was to apply the brakes, it must have been like skating on ice he lost all control of the car and side swiped us right in the door pillar next to me; he then went careering off the road to the left and into a field just missing a tree on the way. He escaped unhurt as did his passenger but the car was a right off. Trouble was so was mine, to fix the door pillar would have cost more than the car was worth so it was written off for insurance purposes. The lad's father was a police inspector up in Paisley and the car belonged to his mother. I got the insurance money no bother and did not lose my, no claims bonus, just so long as I got my excess from the guilty party. That took a while but it came through eventually.

I bought Doris from a Doctor in Ayr and I was going to fit a tow bar when the news came of my transfer. Stuart was kind enough to tow the caravan from the Mull to Girvan for me. All Stuart could say was, "That's what friends are for" at least he accepted the money for petrol even if he wanted nothing for the tow.

Goodbye Mull of Galloway, Goodbye Star of the Four Kingdoms

©Scottish Lighthouse Museum

The paraffin light with the lighthouse pier in the background

Chapter XXIII
Ailsa Craig
The Island & The Station

Trowier Road, Girvan was the address of the shore station for Ailsa Craig and the Block was identical to the ones in Edinburgh except that for some reason they were not painted white, maybe it was so that they would tone in with the drab council houses around them. The house that we were to live in was number 18 instead of 6 but it occupied the same position in the block, so we were more than familiar with the layout. A main difference was that there was a driveway with two garages; the one on the right was ours. Our neighbour above us was Louis Hendry and he was the Principal Keeper on the opposing relief; he would be going out to Rock for the last time in a couple of months before being transferred to the Isle of Man. It was rumoured that Duncan Leslie would be replacing him, in these troubled times it was best just to count everything as rumour until it happened but this time the rumour happened to be true. My neighbour next door was my Principal Keeper and he was out at the Rock when we arrived at the Station; Donald and Greta Michael had two grown up boys but we only ever saw one and that was Louis, they had been here for about a year and Donald would be home in about a week's time. Above Donald were Bob Hepburn and his wife Liz, their grown up son George still lived at home. The other Keepers who made up the relief stayed out at Turnberry Lighthouse and that was where the helicopter landed and made the reliefs from. They were our old friends from the Crescent, John and Carol Douglas with the boys. Not related in any way were their neighbours Norman Douglas and his wife.

I was settling in just fine when Donald came home and introduced himself, I'm sure he was a little standoffish at first but I can well understand why when I heard that he could hear …The Galloway Flasher… from out on the Craig and was not sure what to expect when he met me. Once I had spent a month out on the Craig and he had ironed out a few creases in my character we became a good team and I hope good friends.

Donald without doubt was the best Principal Keeper I had ever worked with and he was a man you could not help but like as well as admire. He is so completely unassuming yet the work gets done. I don't think he would have accepted fools lightly but he was always willing to help and share his knowledge. Most of all he had a remarkable sense of humour that was in built. He would come out with the funniest of quips and remarks with such a deadpan expression that until you got to know him you would take for serious comment. But above all he was a man of his word and he was in no way a braggart; if he said he could do something or he had done something take his word for it.

Donald and I would be picked up by Taxi and taken out to Turnberry, John Douglas would be the man ashore and he would help us on the relief day; he and Norman would have their relief in a fortnight. They each would get an extra days pay for helping with the reliefs.

Donald liked to get a paper from shop in the main street, so got the taxi to stop on the way. We had to wear our survival suits travelling as there was not enough time to change at Turnberry before the chopper arrived. If the chopper came from Oban that was usually the case then they would notify Turnberry of the choppers departure, on the other hand if the chopper had to refuel elsewhere there would be little or no warning of its arrival.

Initially both Donald and I wore our uniforms under our survival suit, but in the summer the heat was unbearable to we took to wearing casual clothes; Donald was a bit concerned on one relief when I seemed to attract a few dogs that kept sniffing at my heels, outside the shop. He thought I was on heat and looked at me strangely, I burst out laughing and had to tell him that I had farted and for some reason it was heavier than air and settled at the bottom of each leg; casually escaping through the vents at my heels. He laughed too but then stuck to the original idea that I was on heat and told Norman to watch himself once we got out to the Rock.

By the way the next time we went out I noticed that Donald had put tape over the vents until we got to Turnberry.

Ailsa Craig is about eight miles from Girvan and about the same from Turnberry, it sits like a dumpling out in the Firth of Clyde and its nickname is Paddy's Milestone. As the crow flies its roughly half way between Glasgow and Belfast but I would argue that it is not half way between Glasgow and Ireland by sea and unless Paddy was a crow I cannot be sure of the nickname's origin.

The big lump of granite that is supposed to be the plug of a long extinct volcano is 1152 feet high; uncannily it shows a different aspect from every direction it is observed from. Looking across from Girvan it is like a Christmas pudding; even more so if it has a dusting of snow or some low cloud is shrouding its top. On clear days I have seen it from Stranraer and admired its grandeur silhouetted beyond the head of Loch Ryan. I have seen it on from the A77 on the approach to Monkton roundabout, maybe one day I will get to see it from Arran and from the Kintyre peninsular to complete the aspect profile.

Looking at the Rock from the cabin of the chopper you are concentrating more on landing than you are at the scenery as the Rock looms larger and larger with each mile. It only takes about eight minutes from Turnberry to the Craig and from Turnberry you cannot see the Lighthouse very clearly with the naked eye; it only has a stubby little tower and the other buildings are low profile too, you can just make them out from Girvan beach but only if you know what you are looking for.

The Station occupies a spit of land that comes out from the Rock like a tail that was formed by a myriad of millions of granite rocks and boulders strewn all around the base; some by nature and some by the workings of man. The granite is still popular especially for making curling stones, although it has been many years since the heyday when there was a whole community of quarrymen living or commuting to the island and it even had its own postage stamp.

The Island outside of the Station is owned by the Marquis of Ailsa so we had to respect that fact and regard him and his heirs as absent neighbours, but the landing, the pier and the helipad as well as the Station were the responsibility of the Board.

Yes we did have the odd visitor and officially we had to check with the factor for the estate when anyone other than Lighthouse people landed on the rock, but we could not report what we unofficially did not see, just as long as they were not up to any mischief and had the courtesy to announce their presence; they came and went unobserved. Besides, anything of any value was at the Station, technically all the granite on the island belonged to the quarry company that had the license granted to them, but if anyone could find one of the rare granite cores that were left and had the strength to carry it to the landing then he would be a guy not to argue with.

Landing on the helipad took a bit of skill especially if there was a bit of wind, having a windsock was useless because you could get contrary gusts that swirled around the rock at the helipad and were different to those at the flagpole; so it was down to the pilot to land us safely.

Our landing pad was an ideal testing place for the Navy Pilots of HMS Gannet Prestwick to hone their skills, their Sea King helicopters were much too large for them to land fully within the pad; they could only land successfully if they put the tail wheel on our towpath. To show that they had mastered the task they would collect a stone and take it back to base. So if you have the opportunity to visit RNAS Prestwick and you see the white painted stones bordering the buildings you'll know where they came from and how they got there.

The Marquise's factor put a stop to the Navy's training; perhaps it was done under the advice given by one Bernard Zonfarillo who represented the RSPB. Any argument he presented that the noise disturbed the birds nesting was shot down in flames because there was no way he was going to prevent the NLB from carrying out their duties on relief days and secondly the birds nesting on the island were well away from the Station; any birds stupid enough to lay eggs on the spit of land would be subject to heat generated by the sun on the granite that would be enough to cook them; or else be eaten by the rats that infested the spit and the lower slopes of the rock. I think that you have gathered by now that I did not think much of the bird man of Ailsa Craig, yes he was probably doing what he thought was right in protecting the birds but he said things in the most obnoxious and self opinionated way that made me take an instant dislike to him.

We did have the pleasure of the company of RSPB researchers who came to catch and ring storm petrels on their migration; I noted that Bernard was not among their number; so assumed rightly or wrongly that he pissed them off also. We enjoyed their company and both Donald and Norman helped them in their night time escapade. I was going to help but remembered that I was as blind as a bat at night as well as being as good as deaf in one ear; especially when I tripped among the rocks as they played now you see me now you don't in the rays of the passing beam.

The Station had boundaries of walls and two courtyards that made a square; the stubby tower stood at the end of the engine room block and the stairwell was accessed through the engine room. There was a block of out buildings that stretched from the tower to the courtyard perimeter wall and the accommodation block formed the central partition, likewise some more out buildings formed the perimeter of the second courtyard. The Station was elevated to about fifty feet above sea level so there was a bogey and pulley system on rails that allowed us to bring heavier stores up from the landing, the winch house was just outside the courtyard at the rear of the out buildings. Now I believe that the quarry company used a similar system to transport the stones from the far reaches of where it was quarried to a sorting shed and then down to the pier. The remnants of track led in both directions North and South towards the disused foghorn towers at either end of the Island. Some of the pipe work could still be seen rusting away after such a long time exposed and uncared for. The towers were the same shape as coconut pyramids except they were concrete and instead of a cherry they had the foghorn coming out of the top.

Up until just recently when foghorns were discontinued throughout the service, there was a typhoon system in operation. The assembly still stood with its array of trumpets and like trumpets they were made of brass and so had some scrap value; during the next couple of months we would get a works order to dismantle it to salvage the metal. I had never heard the system in operation and maybe if the weather was the same as it was at Prestwick the only time it was sounded was on a Monday morning to test it. That is

Prestwick airports major asset; it rarely ever gets fog bound. All I can do is imagine the sound coming from the seaward facing trumpets as it reverberated round the Craig, its pitch much higher that its two predecessors, I could imagine too, the back echo as it bounced off the solid mass of granite behind it; the overall cacophony only heard by those living on the Craig or those unfortunate not to heed the earlier monotones at four miles distance. I could well imagine those haunting tones being the last ones heard from a vessel as she ran aground during thick fog; but as there is no record of such a vessel in recent history, then it would hardly make a good script for a film based on fact. If they can dream up a ghost ship then why not a ghost lighthouse, with ghost keepers to go with it after all in years to come that's what we'll all be, ghosts of the past.

The Light of Ailsa Craig was paraffin but that was where the similarities between it and InchKeith ended. I think the light was built for the little people because what was inside the stubby little tower was miniature in comparison to other lights I had seen don't get me wrong, it was only the lenses and carrier that were small, the lamp was the same size and so was the rest. But only someone severely vertically challenged could stand upright inside, the rest of us mortals and that included all of us keepers; had to do all of the maintenance and lighting from outside the carrier. Given that Duncan Leslie was over six foot tall and Bob Hepburn not much shorter I could well imagine them being able to extend their arm from one side to the other. The Light had the same nominal range as InchKeith's but could only be seen from three of the four cardinals and all points in between anything on the Western side would be obscured by the mass of the Craig.

You cannot say you've been to Ailsa Craig unless you've been to the summit, on a summer's day there was nothing more stamina sapping than scaling to the top nor more invigorating once you got there. You did not need any special equipment just strong legs, for most of the way there was a well-trodden path the traversed its way up, about a third of the way up and plainly visible from the Station was the Monks Well, it was said that in medieval times the Craig was an outpost of an abbey just like the one on Holy Isle off the coast of Arran and the Monks had dug a trapment to catch the water that trickled down from the top. We had a similar set up farther round to the north and ours was at a lower level and collected into a large tank, every now and then we would check to see there were no obstructions. It was not spring water but water that was collected in a peat bog near the top of the Craig, the peat not only acted like a sponge that could absorb thousands of gallons of water in its depths but filter it at the same time. There was no hardness to this water and it tasted sweet, this was the true Uisge Beah...Water of Life, (Whisky). There was one thing that we needed to watch out for that could spoil its sweetness and do us real harm at the same time and that was the rats if they got into the supply, then it would be over for the keepers of the Craig; every so often we would take a sample just to have it tested.

Once you got past the Monks Well then the hike got a little steeper until finally there were some larger rocks that you needed to have hand holds in order to scale; from then on to the top you were faced with lush grass, there was even what looked like a sheepfold; but it was not sheep that were out on the Island at one time but goats. The old Marquis brought them out here for sport but lost interest in them and they were left to become feral it was left to Lady Ailsa to send the factor and a party out to shoot the lot of them because not only had they become feral but the in breading had caused a lot of abnormalities. When I heard that, I was beginning to imagine lots of kids running around wearing lighthouse keeper's hats. Donald had seen the wheels of my mind at work and corrected my train of thought.

There were only rabbits to feed on this lush grass now but they were not as big as you might imagine, their numbers had been decimated by the plague of mixamatosis that engulfed the rabbit population of the whole country. The survivors tended to be on the small side and did not have much meat to them. There was just one steep grassy knoll to the top to the magnificent view that made the exertions all worth it. The whole of the Clyde and beyond opened up before you, 360 degrees of beautiful scenery. From Northern Ireland to the South and the Mull of Kintyre then moving westward across the middle of the Peninsular to the Pap's of Jura and Hills of Islay beyond. Moving northwards in the distance you could just make out the peaks of the Argyle highlands, the Trossach's and the Campsie Fells. In the foreground the Isle of Arran and the lofty peak of Goatfell, the Isle of Bute with its hills and the smaller islands of Cumbrae and Little Cumbrae, then just beyond Cumbrae, the mouth of the Clyde at Gourock. The Hills of Renfrewshire and North Ayrshire conceal the metropolis of Glasgow that lies in the plain of the Clyde. The Ayrshire coast and the rolling hills beyond giving way to the loftier Galloway Hills as we track back towards the South once more. Not to record the magnificence of the scenery but to mark a peculiarity at the top there is a trig point that denotes that it lies at a cross roads of grid lines.

That importance has significance to Ham radio operators, the fraternity that Donald hoped to join once he passed his tests. I hope I cause no offence to either them or Donald if I call them Hams but we had three old Hams out with us for a few days to try and speak to as many people as they could, one of them went to the top of the Craig in order to record as many as he could from each of the different sectors, I believe it was all done for a charity event so I hoped that they achieved their goals and more besides.

There was only one more thing for me to do and that was to circumnavigate the Craig on foot. That was an achievement that could only be done at low water and preferably outside of nesting season; not because you might disturb the birds more like the other way round and fatally so. The young gannets were left alone by their parents to feed themselves but they were hardly fledged before that happened and consequently they had to launch themselves off the 900 foot cliff face to reach the sea below, some misjudged the leap or were not strong enough so landed with a crash on the rocks below, their corpses littering what passed for a beach. Even with a safety helmet on I think a beak would able to penetrate its protection from such a height. Besides if the Gannets didn't get you then the droppings from the gulls or the diving of the terns would. There really was only one section of the Craig that was impassable when the tide was in but you could get caught and isolated if you misjudged when slack water was, it would normally only take about an hour or so at the most but it was rough going and sore on the ankles leaping from rock to rock like a man on hot coals. Talking about hot coals in the summer the granite stones would become unbearably hot and every so often they would give a crack, if you threw them against each other puffs of smoke could be seen, they say that this is radon but it is relatively harmless in the small quantities that we were likely to absorb. In winter it was the opposite, one of the old ways of quarrying for granite was to drill holes and let water expand in them when it froze, doing the job overnight without the need for chisels or wedges. It might also go a long way to explaining why the old foghorns were abandoned, at any moment they faced destruction from hundreds of tons of granite cascading in blocks from the face of the Craig as water had found its way into a crack and expanded. That moment never arrived as I believe they still remain intact, nevertheless it was a chance the Board could not afford to take, and it was not so much the loss of the towers but the loss of the keepers or artificers that may be in attendance at the time that concerned them. All quarrying was restricted to areas to the north and south away from the lighthouse.

The Craig had its partner in the Island of Staffa both had similar features and were made of granite and their volcanic past could be traced in the tall hexagonal columns. The tops of similar columns can be seen at the Giants Causeway in Northern Ireland.

© Vanessa Langley
South Fog Signal Tower

Chapter XXIV
All ashore who are going ashore

There is nothing worse than having to suffer toothache and not being able to do much about it, worst still if you develop an abscess at the same time. With the best will in the world you cannot guarantee that good dental hygiene will prevent you from suffering from the dreaded scourge.

I had the misfortune to still have two milk teeth incisors till I was 21, the misfortune about it was that the adult teeth had become impacted behind them; they grew down and then along the front until they finally met behind one of my two front teeth.

Apart from the excruciating pain as they trapped a nerve in the process of meeting, they would be useless and in the future cause me to eventually lose all of my teeth at the front. The impacted teeth should really have been picked up when I was in the Army. I had to have a couple of fillings but had they taken a full mouth X-ray, then the problem would have been detected earlier. There's an old saying in Scotland…If if's and and's were pots and pans there would be no need for Tinkers…That being so I had to get on with it. The milk teeth were still strong and healthy, so they left them where they were; but I had to go into hospital to have the others taken out. I was left with a couple of pits and it was those pits that caused the problems later on.

The worst time to take any illness is during a spot of bad weather, I'd been suffering with toothache for a couple of days but felt that it was manageable and at one stage it felt as if it might be going away, but like the weather it was the lull before the storm.

It was obvious now that I had developed an abscess and as the pit began to fill, my face began to swell, and the risk of septicaemia grew with each day that passed. We did have a boatman Alec Ingram but his boat the Leonora was laid up having her engine overhauled; that was supposing the weather would let up to allow him out. There was another option and that was to ask Mark McCrindle if he could come out with his boat the MV Glorious when the weather eased, and that was the plan just so long as my condition did not deteriorate, otherwise our friends from Air Sea rescue would be making a genuine call to Ailsa Craig, despite the protestations of Bernard Zonfarillo. Had a weather window not opened up it would just have been a matter of hours before it became imperative that I get hospital treatment. My temperature was rising and it was becoming harder to control, I also had flu like symptoms and generally I felt rotten. We did take advisement from the local GP and there was some penicillin at the Station but it was not working as well as we hoped.

Mark arrived and I was taken ashore, my first port of call was to the doctor who was waiting for me, he had to assess whether I needed to go for emergency treatment or whether he could do anything to lower my temperature and boost the antibiotics. He was successful in his treatment and a couple of days later I was fit enough to go to the dentist and have the pit drained.

We saw Mark out at the Craig almost as often as we saw Alec Ingram on his official fortnightly provisions trip. Mark had creels just off the Craig and if he was not setting them he was taking fishing parties or trippers out for a sail around the Craig. Both Alec and Mark were experienced boat handlers; Alec used to be Coxswain of the Girvan Lifeboat, so the weather had to be really bad before Alec would refuse to put his passengers his boat or himself in danger. I remember going back out after recovering from my abscess ordeal and although the wind had abated the swell tossed the Leonora about like a twig. Not once did I fear I was in any danger; I was concentrating more on keeping my breakfast down. Every time Alec mentioned the word trough as a word of warning, I could not help but think of pigs and swill till I eventually had to position myself nearer the door should I be overwhelmed by the need for Hughie and Ralf. My sea legs were just fine it was my gimballed stomach that was churning and turning me paler shade of green. The eight-mile crossing took much longer than the previous trip on the Glorious but Mark was lucky and I was luckier that the storm had abated long enough for the swell to diminish even when he passed out of the lee if the island into more open water, I could even stand on the open deck without fear of being soaked with spray. But I was confined to the cabin and fighting hard not to mess up the floor; eventually we found the calmer waters of the lee and I was saved any embarrassment. Donald was waiting on the pier ready to take the bow line; he took one look at me and said, "You look worse than when you went ashore." I knew that he was exaggerating, when I left my face was all swollen on one side and it looked as if I had gone a few rounds with Mohamed Ali. Now the boat had stopped and was on a more even keel I felt much better in myself, but Donald must have seen the greenness about my gills. Donald was an ex Marine Commando, so I believe he would have had to cope with a lot worse. Just like my father he said little about his life in the service, but I bet he had a lot more exploits and tales to tell if he chose to tell them.

It was Donald's life in the Marines that caused him to have to go ashore; Donald had contracted malaria on a trip to the Far East, well at least I think it was the Far East, Donald never said. The trouble is some people once they get it carry it with them and it never fully leaves them, so they can get recurring bouts every so often. You could see that Donald was not well, but only Donald knew what it felt like and how bad he really was. All Norman and I could do was to reassure him that the Station could run just as well without him, but ultimately the decision was his. The weather was good, so in the end Donald saw good reason that as he was not getting any better it would be wise for him to go ashore. It was just a couple of days to the Half relief when John was due to come out and Norman go ashore, I don't think Donald would have lasted till then and once the Board were informed of the need to send out a relief for Donald they would have insisted he go ashore immediately anyway. He knew that and that was why he called Alec to come out to take him ashore.

John came out with a relieving Keeper; reliefs were always a fun time at the Craig, especially with just two on the rock. It was easy enough loading the provision and personal stuff on the bogey and attaching the cable to it but getting the bogey moving while at the same time manning the winch house brake was not quite so easy. Norman was dressed ready to go ashore in his survival suit so it would have been a risky business for him to pull out enough cable to take the bogey to the incline where gravity would take over. He needed to man the brake to stop the momentum of the bogey level with the helipad, so it was left to me to push the heavy bogey.

Twice a year we would have to bring up coal from the pier to feed the Station's fires. It was hard work off loading the bags of coal from the boat then loading them onto the bogey. The easiest job would be in the winch house, the winch man would only have to lift the bags of coal once, the middleman and that usually happened to be me would have to lift the bags of coal three times and because of safety reasons did not even have the luxury of getting a ride on the bogey on the way up. I was winch man once and the hardest part about the job was making sure that the bogey did not crash onto the end of the pier. All you need to do was let the bogey run at a steady pace applying the brake every so often to slow the bogey. When the cable came to about forty feet of the buffers it had a yellow marker painted on it, all you had to do was apply the brake and it would stop before it crashed into them. The most unforgivable thing would be to forget to connect the cable to the bogey at the top of the hill. The water at the end of the pier was much too murky to see if anybody in the past had made that cardinal error. I don't think there was anybody in the service would freely admit to it anyway; but I was party to seeing a near miss when a Relieving Keeper who shall remain nameless did not apply that brake soon enough to stop the bogey slamming into the buffers. Thankfully he was only practicing, there was nobody down at the pier at the time so as he learned by his mistake and there was no damage to the bogey it might have remained unnoticed except for the look John gave me when he noticed the impact abrasions on the buffers. I'm sure Donald would have had something to say had he been Principal at the time.

Now getting back to Donald who had to go ashore, John was now Keeper in charge and the Station was in safe hands for the next two weeks. We would soon have to get used to seeing strangers every month on the Craig, Norman was taking up the position of Monitor at the newly established monitoring Station set up at 84 George Street. All of the Relieving Keepers would come from Oban, Edinburgh or other Lighthouses that had empty dwellings. So forgive me all those Relieving Keepers who I have failed to mention in this account, the trouble was of placing who was where at what time. There is an old saying that things have a habit of coming in threes, well I'm not sure what happened after me but just a few days before I was due to go ashore at the next relief, I had an accident. Now I'm not sure who was to blame and perhaps on reflection it might have been a bit of both parties, however, it could have been avoided but it happened. John and I were down at the pier helping Mark unloose the mooring lines on the Glorious; I was at the stern rope when one of the guys on board pulled on the rope to bring the stern in. In doing so he trapped my fingers between the post and the rope, I heard the click and so did John, it sounded just like someone cracking their knuckles. The pain I felt in my fingers and the colour as they turned from a healthy pink to blue should have brought forth the same change in the atmosphere around me but I was stunned more than in shock at the whole thing. Once he had let go his grip and the pressure was released the pain came forth in waves as the blood returned once more to my fingers, the pain in my pinkie remained constant and did not go away like it did with the rest of my digits. After a couple of minutes John asked me if I was OK, I told him that I still had pain in my little finger, quick as a flash he made up his mind to hold Mark there just in case I needed to go ashore.

Mark did not mind a bit, he was at the wheel when it happened but did see the guy pulling on the rope and knew from my gestures that all was not well. As I say it was a silly accident and had I been ashore I would have just have strapped the fingers together and got on with things. Like all employers the NLB have liability insurance but that would become void if they allowed a person with an injury that could incapacitate them in any way from remaining on the rock any longer than was necessary without treatment and being passed as fit by a doctor. First of all John would need to notify the Secretary of the situation and secondly seek advice from a doctor. It did not take long for the doctor to say that he could not make a judgment until he saw the injury, that was it I was going ashore yet again before the relief. The disappointing thing was that once the doctor saw me, his advice was to do exactly what I had already done, bind my pinkie to the finger next to it to immobilise it and it should heal itself. I was not even sent for an X-Ray but I suppose that was understandable, but I could not help feeling that I had let the side down over such a small matter.

© Vanessa Langley
How the Station looks today with the solar panels on the roof

Chapter XXV
Adventures with Donald

I still kept my CB and like Donald used it at about seven at night to speak to Mags, because of the proximity of the two aerials there would have been a good chance of bleed over, so Mags and I would listen out on Donald's frequency and wait till they had finished. It was like listening in to the Archer's and every bit as entertaining, no wonder all the other channels seemed to go quiet about that time every night, Lamplighter had us all listening in. Everyone knew that Donald was talking to his wife so had the courtesy not to interrupt anyway once you got listening to them you were hooked. Sometimes it was the mundane things that he had a way of making funny or entertaining, the things they would talk about were in no way private or personal; after all we still had the telephone for that. I'll give you an example; I told you that Donald was studying up to get his Amateur License, well while he was at home and while he was away he would study Morse code on a practical basis using course material broadcast on the radio. At home he had a loveable old spaniel, a canary and a new member of the pet clan in the shape of a cockatiel. The canary was a good singer but kept his singing down to a minimum until the cockatiel arrived on the scene, he would also go his dinger every time he heard the Morse code. Donald was so distracted by his singing and got so fed up that he shouted at the poor bird; who henceforth never sang another note. Greta was oblivious as to why it stopped singing and I was sworn to secrecy when Donald told me why, but as that was twenty years ago I think it is safe for me to release this official secret. The conversation drifted round to the topic of the birds, the cockatiel was mimicking and singing but the canary sang not a note. She says to Donald, "It was such a lovely singer when we bought it, maybe it's a duff." To which Donald would reply, "No Greta it's not a duff." "But Donald it does not sing, why is it not a duff." "Well Greta my love, if you wake up one morning and see it lying at the bottom of the cage with its tiny feet in the air; you'll know to get out because there's a gas leak." There were many other instances of his humour that were broadcast for the benefit of the listeners and one day I had a compliment passed onto me by the master himself; for once he was in listening mode and was listening to the same channel as myself, I should not really be ungracious towards these breakers but they made the cardinal sin of degrading a chap over the airwaves. The conversation went something like this, "well you know what his house is like; it's never cleaned from one day to the next"..." Yes I know, I went in there once and the smell of sweaty feet, you would wonder if he washed at all"..." Finding anywhere to sit was impossible among the rubbish and that was if you could find a seat at all". By this time I'd heard enough and had enough, they needed to have a sharp reminder that other people could be listening and there were laws of slander, so I chips in unannounced and says "Yes he gets his furniture courtesy of Ivan Boardley." It served the purpose and after a brief round of "who said that?" The airwaves went quiet on that frequency. My reference to Ivan Boardley was meant to be absurd, he was a fish merchant and all his boxes had his brand name on them, his name was well known around the southwest of Scotland so they would undoubtedly get the meaning. Donald's SWR needle would have gone off the scale so he would have known it was me even if I disguised my voice.

He ribbed me afterwards by saying that he knew them personally and would tell them it was me but as usual it was said in his deadpan manner so you knew that he was just kidding. I did not give up my quest to clean up the airwaves or mock the afflicted like old Rosebud who was convinced he had hit skip when I put on a real Basildon come London accent reminiscent of my youth and was answering myself in a different voice. Like April Fool's day I did it in such a way that most people would know it was a wind up, I gave myself one handle called ...Rivet... when asked what my location was I said hanging on the side of a ship going up the Thames, when that did not work I gave the other the handle... Muck Spreader... and I was ploughing a field in the Blackwater estuary. When I was still taken seriously, I made as if the skip had vanished as soon as it came and disappeared off the airwaves but not before I heard Rosebud going on about his encounter with the skip. Donald thought I was cruel to treat an elderly gentleman in such a fashion, this time he succeeded in fooling me as to his seriousness either that or it was my own conscience.

Talking of conscience, morning coffee was a time when the lads would come in and the radio would still be on. I would listen to it while I was doing the housework just like thousands of housewives around the nation, but as the station I was listening to was Radio West Sound only the people of within the range of the signal could hear the dulcet tones of Lou Grant.

On a Thursday he used to have a phone in where listeners could express their opinions on this and that. At various times in his broadcasts he would also have quizzes; what got my goat was he would always assume that his listeners had nothing better to do and were tuned to his every word, so more often than not some answers were repeated by those people like me who were doing the housework. He always rebuked them in such a trite manner that really needed to be tamed and one day I was given just the ammunition to do it. Donald was the gun. I mentioned my concerns to Donald and he said, "don't tell me, tell him", pointing to the radio and the telephone; he was quite right if I was so annoyed by it I should phone him and tell him. Thursday morning was just the time to do it. I could not back down even if I wanted to; all it took was the courage of my convictions and to use my words wisely. I lifted the phone and dialled the number; my heart was racing a bit but I hoped I was about to give the performance of my life. A lassie answered and I was kept on hold until a convenient moment and then I would be straight through to Lou. Lou made the biggest mistake by making reference to Bernard Zonfarillo, the Bird Man, by way of introduction asking if he was out here so my conversation went like this…"Lou there has not been anybody out here all month the weather has been too bad", it was howling a gale and it was pissing down with rain just as I was sure it was doing outside the studio in Ayr, if he had taken the time to look…" Oh I'm sorry to hear that", he was just about to change the subject onto something else when I interrupted and let loose the reason for my call. "Lou ," I said; "I'm calling to tell you that should be a little more gracious towards your listeners, after all they, unlike you, do not have the answer to the questions you pose, printed neatly on a card in front of them and if like me they have other things to do that takes them away from the radio they can miss some of the answers," I was allowed to continue, "Lou before you condemn your listeners, you must first make sure that the answers you are given are the right ones," I then went on to tell him that the answer to who are the governing authority for Lighthouses in Britain was not Trinity House and that Ireland and Scotland have their own. I finished my rebuke by saying; "I listen to you program each morning not only as a penance but also as a reminder that there is someone out there worse than I am." I was not there in the Studio so I cannot say for sure how much my telephone call affected Lou. What I did hope was that Lou would moderate his behaviour towards his listeners, I had that much confirmed when his attitude did change over the next few weeks and so did the format of the show. More than that, his very next caller tried to come to his defence and probably read the same source of misinformation as his researcher by saying, "It say here Trinity House," Good on Yer Lou for replying in more moderate tones, "Let's give Peter the benefit of the doubt that he should Know who he is working for." I was still a bit shaky, when I came off the phone but both Donald and Norman shook my hand saying they did not know I had it in me.

I was brought down to earth with a clatter when Donald suggested that I might want to consider moving to escape the ardent Lou Grant fans that lived in Girvan. To keep the story alive, Donald placed a card in his front window that said Peter lives next door.

You can be wondering about minding your own business when all of a sudden something strange and inexplicable catches your eye, well that's what happened to me one afternoon when I was about to go for a walk; I had just left the courtyard when my eye caught sight of this balloon like object bounding along the spit towards me. My first thoughts were, "I must let Donald and Norman see this", what it could be was irrelevant at this moment because as it was almost as big as a man, I would have problems trying to capture it let alone secure it for closer inspection. I dashed back to call Donald, Norman was on watch so he was stoking up the fire and would hear me telling Donald anyway. We all dashed out, this spherical object was heading straight for us but then took a bounce off a larger boulder and veered off at a tangent as it was caught by a contrary gust.

I am a great believer in discretion being the better part of valour and I had enough of having to go ashore before the end of my stint, knowing my luck, I would chase after this, whatever it was, fall and break my ankle. Besides the last time I saw anything like that was on television chasing after Patrick McGooghan in the Prisoner. Donald and Norman were having a ball or should I say trying to have a ball, every time they seemed to have it within their grasp it eluded them. I had visions of one of them being tangoed if it suddenly burst on the jagged granite, but it seemed more rigid than your average balloon and as yet I could not see where the air went in. In the end a great gust of wind took it higher beyond their reach and it was gone to torment some other guys further up the Firth of Clyde. Donald telephoned the Coast Guard to let them know roughly where it was heading; they might also be able to shed some light as to its origins. As no one had reported anything like it missing, the best they could come up with was a weather balloon. We had already thought on that but this thing defied that kind of logic, aren't balloons meant to go up and stay up. It puzzled us trying to think what it use it would have, I don't think I was remotely close when I thought of it as belonging to an escaped Orca who decided to take his favourite toy with him.

There had been reports of sightings of harbour dolphins, porpoises and pilot whales, we were often visited by basking sharks; I could well imagine a school of dolphins playing their own version of volley ball but when it came to the basking shark it took on a sinister tone as its great mouth vacuumed up the oversized ball as well as the plankton.

After the dishes were done at lunchtime, we would normally have a coffee and then disappear across the courtyard for a game of pool or snooker on the half size table. Sometimes we would depart sooner than planned if the wind suddenly blew across the H can forcing smoke down the chimney, it was worse if the fire had just recently been charged. Donald would sometimes brave the smoke and say to us; "you think that is bad, I remember once when you could not see your hand in front of your face, it was that thick." We used to call the afternoon watch…Bunkers… not surprisingly because that was generally when the fire was cleaned out and relit, one afternoon it was my watch and we had a blowback; but instead of the usual puff of smoke this time it was belching the acrid sulphur type stuff you get from a recently lit fire. We all had to evacuate the living room, but it was my job to find out the reason why. Donald's room was just across the lobby from the living room, so I shut the living room door and then dowsed the fire to put it out. Initially I would be generating more smoke, but I had nothing to lose anyway because the smoke was still belching out thicker than ever. Passed experience had taught me that it must be the H can, and sure enough even from ground level I could see that it was tilted at a queer angle. The problem was that it was blowing a near gale and it was raining, so trying to fix it was not going to be easy. As it turns out the can was useless and could not be fixed, we kept a spare can on Station just for this eventuality. When Donald saw me on the roof he was almost insistent that I leave it until the wind calmed down a bit, trouble was we had a back boiler and needed the fire for hot water and secondly there was no guarantee that the wind would get any better any time soon. Weighing up the options and the fact that I was half way through and wet enough for it not to matter, I persuade Donald to allow me to finish as I felt safe enough and could manage the awkward weight of the can on my own. Getting the can back on was a bit of a struggle and before it would sit in place, I had to wire brush the deposits of soot that had collected. That was when I decided to go the whole hog and sweep the chimney at the same time. I thought to myself there was no point in going through the process of fitting a new can and then having to refit it a couple of weeks later when it came time to sweep the chimney. Donald thought I was going out of my way to fix the problem but admired my fortitude none the less. By the time I was finished, I was soaked through but the can was secure and I could get the benefit of a hot bath and sit by a roaring fire; I was on cook's duty that week but Donald was adamant that I had done enough for one day and I could clean up any smoke damage in the morning. I did ask him if the blow back was anything like the worst he had experienced, "No Peter I can safely say that I never want to see that kind again." He got his wish, and the next time the H can came adrift it was not me on Bunkers although I did help the other keeper fix it and at least it was a dry day.

Rats around the Station were not really a problem unless their activities became more prominent and the sightings more frequent. The worst fear was of them getting into the water supply but the tank was checked regularly and it was outside the Station perimeter. The rats had acquired a taste for the soft soap that was kept in a drum in the washhouse. In the washhouse there was also a boiler that was great for Dhobying… an old navy word for laundry cleaning… especially towel and tea cloths; we rarely ever used the soft soap unless we were washing the engine room floor, but as the only engines in use now were the generators there was seldom need to wash the whole floor. I was about to go into the washhouse when I saw this rat come scurrying out; Norman had been had moved to pastures new and I can't for the life of me remember what one of the many the Relieving Keeper's was out at the time; but both he and Donald were sitting in the living room at the time. I say to Donald, "We've got our visitors come back in the washhouse again." It must be like some kind of addiction but once they get a taste of the soft soap they can't stay away from it. We knew they would be back, it was just a matter of time; sure enough they for indeed there was two of them I saw sneaking back into the washhouse from the kitchen window. This time I would have an accomplice to help me dispatch these two brazen rodent junkies, I had already prepared what I thought were adequate weapons for the job in the form of a pair of yard brushes.

As silently as we could we crept towards the open washhouse door preparing to do battle with whatever came out, you would have thought they would not stand a chance against us but they had the one thing that both we Keepers lacked that day and that was wits. "The Wily Wodents were too quick", as Elmer Fud would say and like a pair of Numpties we had to report our lack of success to Donald who just sat in his chair shaking his head and tut-tutting his disdain. Well my pride was hurt enough without Donald rubbing it in, so I say to him, "Well if you think you can do better, go ahead and let's see what you can do." I could just hear Donald muttering something like, "send boys out to do mans work," as he went to put his boots on. Within a half hour the rats were back, this time Donald would be the dispatcher, one of us would chase the rats out into the open while the other blocked off one of the archways with a board, leaving only the other as an avenue of escape. Donald positioned himself there with a spade; we both looked at Donald with puzzlement, wondering what he hoped to achieve with what we thought an unwieldy implement for such a task. We were all ready and about to see the master at work or a man with egg on his face eating humble pie. First one rat ran out and made straight for Donald; like the cleft of a mighty sword Donald's arm swung down, a Kerching resounded round the Craig as the blade struck concrete. To our complete astonishment it was not only the concrete that was struck but the rat as well.

It would have been silly for us to have laid bets on whether Donald could repeat the performance with the other rat, and when the rat ran towards me with serious intent of escaping, I thought he might succeed and we might never get to know. Hot on its heels was the other Keeper who threw a timely broom head at it diverting its course in the opposite direction towards Donald. This time we could watch the master in awe as he dispatched the second rat with the same deftness of hand. Donald this gentle and unassuming man had a side to him that spoke of courage and determination, he would have undoubtedly been a mean adversary but the man to befriend and be with in times of conflict.

The sound of the second Kerching seemed louder than the first and if rats are as intelligent a creature as portrayed it was as good as a notice telling all rats to stay away. For a good while it worked but their addiction to soft soap was stronger, and perhaps the other Keepers weren't quite so rigid in keeping the rats out of the Station. It was not disease that we feared they would bring but if encouraged they might build nests and start making a nuisance by eating their way through anything that might prove a material for nesting in just the same way that mice do. They say that you are never more than seven feet away from a rat, well as long as there are walls between us and I don't see them then they can stay seven feet away.

On a second encounter with a rat Donald's approach would be different because the little blighter was seen bold as brass in the engine room but had scurried under one of the engines, we could not leave him there and we had to try and persuade him to leave if he had not already done so. The trouble was that a ducting channel had been cut in the floor to take the cables for when the Station was automated, so the rat could have made his way from underneath the engine to the channel and then follow the channel into the other outbuildings. We needed something to flush him out; the intention was not to kill him at this stage unless the opportunity presented itself and if he escaped in the process so be it. Donald had the brilliant idea of using ammonia to do the job, we did not have any at the Station but we could borrow a little from Bernard who kept some in his bothy. The bothy was outside the grounds and bore no resemblance to those at Lighthouses. This one had no electricity for a start and if my memory serves me well only had two rooms, if I was going to look in there I would need a torch and hoped that Bernard didn't keep anything sinister than ammonia in bottles. My search was in vain because I could not find any besides the place was a bit of a tip and it was a wonder where to start. Donald was one up on me because he had seen it on two occasions and had some idea where to look.

Donald poured some of the liquid into the channel at one end while I waited at the other blocking off the way back into the engine room. He wafted some air along the channel with an old pair of bellows that were at the Station; I waited hoping to see the first wisps of vapour and not a rat, I waited and waited, then heard Donald shout, "Has it come out yet," I was not sure whether he meant the rat or the vapour but as neither had appeared the answer would have been in the negative for both, so I was about to reply taking an intake of breath preparatory of the shout when the fumes caught the back of my throat. My reply was a fit of coughing with a hasty retreat out of the room as the smell of rotten eggs followed me. Donald took one look at me and said, "I take it that's a no then?" We both laughed about it after but I think it was the rat that had the last laugh.

The company that had the quarrying rights came from Wales and every so often they would phone up to see if we had found any cores, any that we found and brought to the pier we would be paid for. It had been a while since we heard from them but we anticipated nothing untoward so we had a good look especially towards the north end where the fog signal was. There were a lot of larger granite blocks there as well a profusion of chippings', all of which could conceal the rare cores that were the beginnings of curling stones. It was one thing finding the cores but a whole different ball game getting them to place where we no longer need to carry them; were they heavy I hear you ask? Put it this way by the time I had finished even my muscles had muscles. Compared to how they looked beforehand my forearms were starting to look like Popeye's. When we got back to the Station after a hard day every bone in my body ached. I don't know how Donald felt because he never said but I think he was suffering every bit as I was. Eventually we unearthed around ninety cores and had them where they could be moved the rest of the way to the pier by wheelbarrow. That would be almost as hard a task in that it would be backbreaking work. Now that things were just about ready we gave the company a call and our jaws hit the floor when we were told that they no longer had the rights to quarry for the stone; that was passed onto a company based in Mauchline Ayrshire, which as the crow flies is just over 30 miles from the Craig. Nothing ventured nothing gained we telephoned the Mauchline company to see if they would pay us for unearthing cores, we were absolutely gob smacked when they said they were not going to pay for anything that was already theirs and that we were to turn over to them any that we found.

There was one thing for sure they were definitely not going to get the benefits of our labour unless they paid for it or earned it the same way we did, even if we had to put back every stone where we had found it. Now there could be a legal argument as to the ownership of the stones and that would be in the fine print of the License to quarry stone, we did not own the stones and it was not our intention to permanently deprive the rightful owners of them, so we could not therefore be charged with the crime of theft. Basically we were as stubborn and as obtuse as they were but it is they who would have the problem of locating the stones not us; and they would only be of use to them if they found them.

Sure enough an old LST...Landing craft...came out to the island, hired by the company for a survey expedition. They had not applied for permission to land at the pier and so that facility was denied them, it was not our belligerence that denied them but their own arrogance in believing they could land where they liked, with the thinking that as they had notified the factor of their intentions that was enough. Donald would have needed to notify the Board to seek the authority to grant access to the pier especially for a commercial enterprise, if the vessel had collided and damaged the pier and the vessel was not insured then the cost would have to be met by the Board, that was the sole reason for writing to the Board to gain access. It was not as if the craft was not made for that kind of landing it was just that the granite being as hard as it was, did not take much pity on the ramp as it came crashing down on the south side of the spit.

I think there was a company director, an engineer as well as a surveyor in the party that came ashore. They had the decency to come to the station to announce their arrival and then they audacity to ask if we had seen any cores lying around; it was too much for Donald to take so he walked away in disgust glancing at me with a prearranged signal for me to take over any further dealings with them; he would be playing snooker. While Donald was still within earshot I said that for a fee I could show them the best place to search and even draw them a map, when they asked how much I came up with the figure £954.60 and I would throw in a map for good measure. They declined my offer and continued to enquire if they could land at the pier in the future; truthfully I told them that they would have to apply to the Board for that. That was the last we saw of them and good riddance; they would find them in the end but it would not be with our help and as for quarrying well that industry would in theory cause more harm to the birds than any helicopter operating on the opposite side of the Craig. Donald asked me when they had left where I got the figure of £954.60 from, I replied, "I reckoned on £10 per stone; £50 for my survey charge and £4.60 for the map". Donald just laughed.

They say that everybody is entitled to their 15 minutes of fame. I had mine on Radio West Sound, but Donald never said a word about his. Can you remember when Channel 4 first started broadcasting; well for a few months prior to that they broadcast test programs. Oddly enough some of the test programs were very entertaining and instructive; so during the afternoons while I was ashore and Mags was getting ready to pick the kids up from school; I would watch some of these programs.

Low and behold, a program came on about a Lighthouse, there was no dialogue to the film but I instantly recognised the Station as Girdleness. I had never been there but had seen pictures of it before; its most distinguishing feature matched the name because about a third of the way up the tower was what looked like a girdle. Not only did I recognise the Station I also recognised the casually dressed keeper who just as casually walked up the pathway to the door of the tower. There was no acting for the camera on Donald's part he was every bit himself and every bit a Lighthouse Keeper; He maybe a Ham Radio Operator but he was certainly no Ham Actor. After he had gone through the process of lighting up, the film came to an end.

I saw Donald out in the garden and went out to speak to him, "You quiet old dog you," I said. Donald looked at me, wondering what had brought on the sudden onslaught of mild euphemism. Quick as a flash, he replied with a few rough barks with his hands before him in the begging position. It was Donald's polite way of saying come to the point Peter or we will be here all day. I then went on to say that I had just seen the film. "Oh that film," he said and quietly went on with his gardening.

Chapter XXVI
Changes on the Rock

Donald and I were not out on the Craig the Christmas a helicopter landed on the top of the Craig on Christmas morning. It was a publicity stunt arranged by a local newspaper and there was a chap dressed in a Santa suit clambering down the pathway to give the Keepers their presents. We found out later that the presents were fakes, just empty boxes covered in wrapping paper. I say to Donald, "What do you think of that then, Donald." He replies, "Well had I been out there we would have all hidden, and then they would have to give the pretend presents to pretend Keepers."

It was inevitable that Donald would probably be transferred before he retired from the service, even more so when he took another bout of malaria, the good news for Donald would be that it would be to Corsewall Point and he would be there until the Station was fully automated. With every day that passed my time in the service was drawing closer to the end, like the sand in an hourglass the last few grains seemed to trickle faster than the first. At that time I tried hard not to think much about the future and seemed more concerned about what was going on now, I suppose in a way it was much like a bereavement that I had not quite come to terms with.

Once I had sorted myself out, I needed to make plans for Mags and the children's sake. I had thought seriously about Social Work, but lacked the confidence and the qualifications needed. First of all I needed to get back into a learning environment but in a small way to start with. I went back to school, literally; while I was ashore I went for a few hours a day to Girvan Academy to study English and Biology the Maths I could do out of a book on the Rock. The following year I passed my SCE 'O' grades in English and Biology, I did fail my maths, and that was a shame because I was hoping that it would have given me a positive inducement to do an Open University diploma in Social Work. It did not deter me and in the end I did a couple of mock tests to see if I was not only capable of doing the course work but had the self-discipline to do it. That I did pass, the rest was down to timing.

We were all saddened to see Donald and Greta leave but just as intrigued as to whom would replace them. Gone were the days when you could safely say that you would get new neighbours, what with travelling Relieving Keepers; we would now be getting a travelling Principal, Jim Hardy had bought a house in preparation for his retirement once the Craig was automated. Since the stopping of the fog signals the pace of automation had gone into high gear. Now nearly all of the Stations had some work done in preparation for automation.

As yet the Craig had only the ground-work done so to speak, all of the out buildings had been tanked so that they would not need maintenance during the many years to come without the loving attention of Keepers, the channels for all the cabling were in place; but all the equipment was still in the process of being tested; especially the solar panels that would keep the batteries charged for the monitoring equipment.

The only time we met was the night before the relief, the Board still liked to see a Principals signature every month on the shore stations returns book even though it was I who was doing all the paperwork. It got to the stage where Duncan would relay the relief information to me to pass onto Jim, because sometimes Jim would not arrive at the shore station till after eight at night, the last thing you wanted to be doing was lighthouse business on your last night ashore. It suited Jim and his family, but that was all; Donald's house remained empty till the Station was de-manned.

It was not really the best start to a working relationship and unfortunately things were going to go from bad to worse. Perhaps my expectations of Jim were too high and in contrast to Donald he was almost the exact opposite. I'm going to be quite frank and call a spade a spade even if it might be the pot calling the kettle black.

If you wanted an actor to play the part of a lighthouse keeper, then he would be every director's dream until he opened his mouth. Without any makeup Jim would look every bit like the traditional portrayal of a Lighthouse Keeper but that was it entirely, Jim had to look the part and act the part, rather than get on with the job of actually being one. Jim had his 15 minutes of fame on TV in the program, What's My Line. The object of the program was to get the panel guessing your occupation from a brief mime and a few well chosen questions, it did not take them long to guess.

Sometimes it was if he had read the script given to him by the NLB but he did not like his lines and decided to ad lib now and again, in the end becoming the worst kind of Ham Actor. These are only my opinions and maybe I never got to know the real Jim Hardy, it is more than possible that he was acting out the part for the benefit of his wife; who it appears wore the trousers in the relationship.

What was to annoy me most was the fact that he seemed to be taking a back seat out at the Rock rather than lead from the front. There seemed to be no direction coming from him at all, it was not really a problem in the day to day running of the station, we were all experienced Keepers and knew our parts; but the direction was lacking when it came to works orders. There was no forward planning, no schedule of work no nothing. Sometimes the only indication we got was when he took the notion to pick up a paintbrush and start working.

It all came to a head during the summer; I had it arranged for Mags and Gavin to come out to the Rock for a visit, this was something sanctioned by the Board and agreed by the Principal Keeper. Once arranged it could only be revoked by the Board if it conflicted with the exigencies of the service at the time, if the weather was bad or if there was something happening at the Station. On the morning of the trip Mags went down to Girvan Harbour with Gavin to meet Alec where the Leonora was tied up. They were not the only ones going aboard to go out to the Craig, Jim's wife and their nephew was going out as well. Mags' did not mind, the company on the trip would be good, especially as she did not like small boats. All that was to change half way across, the lad became so upset that Jim's wife insisted on the boat turning round and going back to Girvan.

I was expecting to see the boat at the landing; not get a phone call from a very upset and distraught Margaret telling me exactly what had happened. I tried desperately not to become enraged the more she went on, the final straw was when there was absolutely no apology whatsoever from Jim's wife, who it appeared as though she every right to commandeer the trip and force the boat to turn round on her say so.

It took me several minutes to compose myself before confronting Jim Hardy, any respect I had for the man had now disappeared entirely. What he and his wife had done was unforgivable and was outside all sense of decency; what's more he compounded it by abjectly refusing to enter the details in the returns and the monthly letter. That would in effect show that the landing took place and would not contradict the bill presented to the Board for Alec's trip in the Leonora. All that did not bother his conscience one bit, for the moment his wife would come out snow white but that was until the shit hit the fan.

With all my respect for the man gone, there was no way that I could continue working with him, I was repulsed by the very look of him; I had to make my feelings known and the reasons why. I wrote a letter of complaint to the Board Stating my grievance with the strong request that I be transferred as soon as possible. By now I only had a couple of years left to go and was given the date of March 1992 for my redundancy. Coincidentally the grape vine had it that there was a shift going to take place at the Rhinns of Islay within the next couple of months and having thought about where I should go if my request was granted; I wanted very much to stay within Strathclyde Region because it would be easier to get a council house within the area we chose to live in when I left the service. Realistically the Rhinns would be the only chance I had of getting that wish without too much bother. I made all this known in my letter and said that this was a position that was forced on me by the actions of one man.

Sandy Wright came out to see me to take a statement and asked if I wanted to take the action further by going to the Union. I saw no reason to involve anybody else if the matter could be resolved otherwise, I am not a vindictive person and the Board could make their own investigation and come to their own conclusions if he had in any way infringed the regulations or had shown conduct detrimental to the service. They could punish him or me for that matter if they thought that I had over reacted, as they thought fit and that would be the end of the matter but what would still remain was the fact that I could not work with that man again.

Chapter XXVII
While you were away at The Craig

In a way I was a bit disappointed that it was only Girvan that we were going to; while we were at the Mull of Galloway and before the bombshell of the foghorns was dropped, I was feeling more adventurous and hoped we might go somewhere like the Point of Ayre on the Isle of Man, I might even meet up with Anne again. I did not feel as if I was coming home even if I was so close to my roots and my family. I was not ready for that yet.

Patna is only nineteen miles away from Girvan, yet it was a train and a bus or two bus rides away that nearly doubled the distance. I would not say I was lazy but I had grown used to being driven everywhere and much preferred the half hour run in the car rather than trail three unruly children on the bus. I am being a little unfair on them but they did pick their times to play up, more often than not when I was at my most vulnerable.

Gavin seemed to know my weak spots and just how far he could go; yet he also knew how to get what he wanted without throwing tantrums. That was not his style; he used his wit and his guile to win people over. Deep down the girls must have thought that he was spoilt rotten; maybe he was because Gavin came along at time of plenty, unlike the girls who arrived during the first years when we were struggling. Whether they felt like that or not they never showed it, well not while I was about. Gavin also worked his charm with them as well; sometimes they would take the blame for something that was his fault just to save him from getting into trouble.

It was usually Jessie and Jennifer at the weekend that came to Girvan. My sister was only five years older than Karen and it was strange in a way watching an Auntie playing with a Niece as if they were best pals.

Just like Edinburgh it was through the children that you got to know someone well enough to call them friend; mine was Margaret McLeod, Norrie her husband was a local copper and they lived in a semi-detached bungalow that was virtually next door. Kirsty went to the same school as their twin girls and Gavin was the same age as Colin their son, more often than not the children would be playing in one house or the other, later once we got to be friendly then that was where you would find us. It was not the same kind of bond that I had with Anne McIver but Margaret was the next best thing.

Girvan was a bit higgledy-piggledy when it came to schools, Karen was at the Academy now and that was the closest of all the schools to the house. Kirsty's primary was in the direction of the harbour but not quite as far; but the infants and the nursery, was right at the end of the town. I'm short enough as it is, traipsing that distance twice every day for three years and I would be worn away to nothing. Peter used to say that was what feet were for and it was shoes that were worn not feet. Then he would complain that I was like Imelda Marcos with the number of pairs of shoes that I had.

What to do about Gavin? Well just across the road from Kirsty's primary School there was a Catholic Primary school, Peter and I had talked things over and we saw no harm in enrolling him at that school. We had no fear that he might become indoctrinated in fact it might help him make informed choices later on in life. At four and five years of age he would have a lot to take on board with going to school so it would just be part of the pattern of his life for a while. He liked his school and got on well with the teachers and accepted that some of them wore costumes instead of normal clothes. He used to call the priest the church manager and had not quite worked out the true nature of his job, the main thing was that he was learning and enjoying learning.

Gavin had not been used to passing shops every morning; he was lucky if he came with me on the weekly shopping trip into Stranraer, when he did, he would always get a toy of some kind. He was moving away from Thomas the Tank engine and was more into He Man and the Masters of the universe, the High Street was the quickest way to get to the school but it also took us past Woolworth's and since that was where he got most of his toys from he thought that he should get a toy every day. He would put it in such a way that it was hard not to say no, it went something like this; "Mum can you get me this model to go with Skelator?" "No Gavin", I replied. "But Mum I can't let Skelator fight He Man on his own." "No Gavin, Maybe next time." "But Mum, Michael says that tomorrow they won't have any." "Who's Michael?" "He works in the shop." It would not be hard to believe that there was a Michael the shop, after all Gavin was a regular visitor; I would be far too embarrassed to go in and ask if Michael was there and I think Gavin knew that. So I was defeated once more and once he had collected all of the models he would go on to others, we did draw the line when he wanted to start collecting Transformers because they would be far too expensive.

Karen and Gavin both had birthdays around Christmas, of course Karen's was on Christmas day so it was even worse for her, but it was hard on them when Peter was out at the Rock on not be there for their special days. They liked to give presents as well as get them but their Dad was not home to see their faces light up at what we had bought them, and they would have to wait till he came home before he could open what they had bought for him. The novelty that surrounds Christmas and birthdays had worn off by then so it was not the same. Peter used to say to the Children that he got to unwrap his Christmas presents every day until they were nearly a year old and his birthday present came in the smallest wrapper.

It was just before Christmas that Gavin had to see the school doctor for an annual checkup. We were sitting outside with another mum and her wee boy when the boy says to Gavin. “Guess what I’m getting from Santa, I’m getting Bed Bugs,” to which Gavin replies. “HuH, I’ve got the real things and I have to shake them off the bed before I can get into it.” I had nowhere to hide my face and wished I could be anywhere else but there. All I could do was try my best to make light of it by shrugging my shoulders and smiling. We were next in so I did not have to keep up the pretence for long. Oh by the way I hope that you were not thinking that there were real bed bugs; it was just Gavin at his best or should I say worst. What followed next when we went in front of yet another stranger; would have turned any mother the colour of beetroot. The doctor says to Gavin, “Lets pull down you trousers, so that I can have a look at your wee Willy” to which Gavin retorted, “Wee! Wee! It’s absolutely huge when I point it.” The room erupted in laughter except of course for me, had I not been his mother I would have thought it funny too; nowadays it is I who torments and embarrasses Gavin by reminding him. At the time he was not to know that women believe that is what all men think about the size of their privates, only for us to know different. After that, I would take Gavin to his appointments but wait outside, next time it was to the dentists and Gavin was on form because within minutes he had them all in hysterics, and I never enquired what he had said. Gavin has kept his sense of humour but thank god he grew out of embarrassing his poor mother in public.

He did have one last throw of the dice when Catherine Leslie kindly offered to give our Karen a computer desk to do her homework on. Catherine had not changed a bit since the Crescent and like me just got a little bit older, I was the only ones with younger children so, like the kind matriarch that she was, anything she thought might be of use she thought of the children. She calls to me in the back garden, “Margaret would Karen like this desk,” It was a nice desk and Karen would be thrilled to bits but I hardly had time to reply when Gavin chirps up, “It’s a nice desk but ya ken she would like a nice lamp to sit on it.” I implored Catherine to ignore him, but too late she shouts out to a bewildered Duncan. “Duncan, Duncan, go up into the loft and see if you can find that old desk lamp.”

Kirsty had a run in with the law when two big burly policemen came after her for running into the back of their police car. The car was sitting outside Norrie’s house and I think they were on a break or had to see him about something. Kirsty was not watching where she was going and went slam into the back of it just as they were coming out of the house. She had never seen Norrie in his uniform so did not recognise him, she had got a shock from the impact but Norrie was going to give her an even bigger fright when he got out his truncheon, handcuffs and notebook and went through the act of arresting her. She was about to burst into tears when Norrie took his helmet off. Mind you he did have a look at the back of the car just to see if there was any damage.

We had sold our caravan to Alan and Pauline our neighbours on the other side; they had boys Stuart, Malcolm and Douglas. Alan was a teacher at the Academy and I often would baby sit Douglas while Pauline went for the other boys, later I would baby sit the pups that Katy there black English setter had. They kept one of the pups and named her Linnhe. Someone had been teaching her how to smile but unlike Kelly's, Linnhe's was a full frontal smile that had us in stitches every time she looked over the fence at us, like her mother she had long curly hair covering her ears that looked every bit like a perm making the whole scene look hilarious and adorable at the same time. They also had a cat called Penny, Penny and Jess did not get on very well but most of the time they ignored each other. Penny had a rather unusual Meow that sounded like the passing of a racing car…MeeeeeeeeOw, but hoarsely; Peter called her Maseratti because of it.

I did manage to get out and see the Craig with Gavin and the girls; Jennifer was with us for a bit so she came out too. I'm not very good in small boats and to me that was what the Leonora was; out in that big sea she looked lost and we had not even left the harbour yet. I looked across at the Lifeboat and thought that even that was small but I would feel a lot safer on her. I was not comforted by the fact that Alec was once the Coxswain. I hoped it was not going to be rough. As it turned out, it was a beautiful day and the sea was almost flat calm.

Peter met us at the pier and Alec shouted what time he would be back for us. It was not like I had expected, I'm glad I had taken Peter's advice about wearing good strong shoes and not sandals, the stones were everywhere and there did not seem to be a bit of flat ground anywhere except the lighthouse. There was not time for us to go to the top, so we had to make do with a walk to the foghorn tower at the south end and the Monks well. It was my ankles that were the sorest and if it was anything like this on the way to the top, I was glad we were only going as far as the Well.

When Peter told me about the well, I had an altogether different picture in mind. I never expected to see a stone tower that was a bit like a turret with stone steps that led half way up, I mean they were just stone steps built into the walls and open. No banister rail no risers no means of support, nothing but steps. I say to Peter, "I hope you don't expect me to climb them!" I know Peter of old; when we went to Frontier Land at Morecambe he had me walking across a conveyor belt type bridge saying that it was the only way in or out of the maze of obstacles. I believed him as well, I tried everything to get off it, plus I had Gavin in my arms at the time. Peter and the girls saw the funny side of it, the more they laughed the angrier I got; finally Peter came to my rescue and lifted both Gavin and me up over the banister. He who laughs last laughs longest; about twenty minutes later, Peter braved something they called a death slide, all it was really was a chute with a shiny wooden floor, the only thing that made it scary was the fact that it was almost vertical for the first ten feet. Peter went down it like a rocket but got friction burns on his elbows that were sore for hours afterwards. Next to that was another one that was even steeper, Peter looked down and walked away. A lad behind him says in his English accent. "What's a matter mate? Lost Yer bottle". Peter replies back to him just as he is disappearing down. "No mate I've got it but I want it to remain intact".

This time Peter was adamant that no one was going to even remotely think about climbing those steps; I noticed that he was looking directly at Gavin when he said it.

©Mearns Forge

It was a boat around the size of this one that Alistair took across to Orsay in Thankfully it had an outboard otherwise we would be still trying to row

Chapter XXVIII
Islay and The Rhinns

Once again it would be Duncan and Catherine who would be bidding us farewell, and Catherine would make us something to eat before we left. This time there would be little chance of us meeting again as the days till my redundancy loomed ever closer.

I was looking forward to seeing Islay again, only this time now that I had the car, I hoped to see all of the island and perhaps even cross the sound of Islay to Jura.

On every other day but a Wednesday, the ferry service sailed from Kennacraig to Port Ellen. The morning service went from Kennacraig to Port Askaig and then continued up the Sound to stop off at Colonsay before arriving at Oban, she would then make the return trip the same way. Our first trip would be via Oban. We didn't have a great deal of choice what day we sailed nor the time of the sailing, it was a five-hour drive from Girvan to Kennacraig on the Kintyre Peninsular and the about same to Oban. As the crow flies it is only about 50 miles from Girvan across the Firth of Clyde crossing the bulk of Arran to Kennacraig; but by road the distance is nearer 200 miles and once Glasgow was behind you and you got onto the A82 at Loch Lomond then the road began its twisty windy route through some of the most scenic country in Scotland. If you have the time you can take the route to Oban via Inveraray it will cost more in petrol too but it is well worth the trip. I had an ulterior motive for taking that route and that was for the most part up till you get to Inveraray the route would be the usual route to Kennacraig. Alternatively you could get the ferry from Gourock to Dunoon and cut out Loch Lomond altogether or if you really felt like a maritime adventure; get the ferry from Ardrossan across to Arran and drive round to Lochranza to get the short ferry crossing to Claonig on Kintyre, from there to Kennacraig its less than ten miles.

Once the novelty wears off and you have made the trip a few times a year it can get pretty tedious and exhausting for the driver, but if you are an Island dweller it is part and parcel of the way of life.

Just as on our previous moves we could not abide by the song …my old Man says follow the van… and we certainly had enough time to dilly dally on the way. The van would arrive at the Shore Station the following day so we would have to stay the night in a B&B. We made it to Oban in time for a late lunch and parked in the queue for the ferry. This would be my first time as a driver on a ferry and I was a bit nervous of making an arse of parking the car on board. Once the ferry docked I had even greater cause, because cars were being loaded by the side ramp and had to reverse into position, the further you were behind the queue the more difficult and less room there was to manoeuvre. Thankfully there were only five cars in front of me and I did not make an arse of myself, unlike the time when Mags' and I went on a bus tour to the Isle of Wight; that was about as close as we got to a honeymoon. To save money we took a packed lunch with us and I had a bottle of orange in a bag. It is only a short crossing, so we stood on deck and rested our bahookies by propping them on our hands behind us and leaned against the superstructure.

The bag with the orange juice swung gently back and forth with the motion between my legs. It had either gotten a puncture or I had failed to seal it properly but it began to leak and a trail of orange liquid began its quest for level ground till it swirled back and forth in an ever-growing puddle unseen below me. The ferry took a little steeper roll that almost made me need step forward for balance and it was then that I noticed the orange running away from me. A bit embarrassed at what others might be thinking I beat a hasty retreat as far away from the scene; trailing a hysterical hyena behind me as well as the trail of the orange until I found a suitable bin to put it in.

The girls enjoyed their first cruise; they stood on deck with me to let the breeze and the sea air blow through our hair as we watched the coastline of Mull fall away to more open water on the starboard side. On the Port side clusters of small islands went from one to the next so quickly that the overall appearance was as if the land was one coastline, only when the ferry passed the Sound of Jura and you were looking back could you see that they were separate islands. The Pap's of Jura concealed their form from this vantage point and look like any other hills and it would not be until we entered the Sound of Islay that all would be revealed; first we had to stop at Colonsay.

There was another good reason to be on this route and that was to see the lighthouse at Ruvaal, where I was a Supernumerary. It would be too arduous a trip to take by small boat or going on a long hike from Bunnahabhainn so the girls would have to make do seeing it from the deck of the ferry.

It was not long after we passed Ruvaal that we had to go down to the car deck ready to get off. There is quite a steep hill to negotiate out of Port Askaig and I was thankful that we had sold the caravan and was just as thankful that we were not stuck behind one; or worse still, the lorry that had kindly parked off the road to allow faster traffic to pass by. Before finding our B&B we intended to have a look at the new house, but as things turned out the B&B come Hotel was on the way into Bowmore and overlooked Loch Indaal, so we signed in and unpacked what we needed for the night, we would still have time to see the house before our evening meal was ready.

Of course I had seen the Shore Station before so knew exactly where it was and what it looked like, Mags' first impressions were good ones and it would be the first time that she would live in anything like a semi detached house. We already knew our neighbours both sides, John and Margaret Bambrick were our next-door neighbours and Ronnie and Jeannie Gould were just over the fence. John would be my opposite number so we would rarely meet, this would be my first time as a half relief Keeper, all that meant was that I would be going out on my own and have two Principal Keepers; one for the first fortnight and the other for the second. Ronnie was out for the first fortnight and Mearns Forge was out for the second. Mearns lived next to Ronnie with his wife Irene and daughters Pamela and Michelle. Pamela was going out with a local lad who ran the garage in Bowmore; they planned to get married the following year. Michelle was in her final years at school and we rarely saw her.

The twins were all grown up now, Alison was in the Army and Fiona worked on board the CalMac ferry the Isle of Mull. Shirley was still a nurse in Edinburgh and Ronnie Jnr was married.

Marion and Karen were older now but it was good that the girls would have familiar faces to play with as for Gavin he would just have to make new friends.

The other two Keepers that made up the reliefs were Bill Crockett and John Carmichael, John and I had met before at Ruvaal so it was nice to see his friendly face again; he lived with his brother not far from the Shore Station. He was one of the fortunate Local Assistants who were retained just to make up the numbers, while there was still a manned Lighthouse on Islay; John would still have a job. Bill lived in Oban and made his way down to Islay, staying overnight before coming out with Mearns.

We had a few days to settle in before I went out to the rock and we would take a rest from moving furniture about and take the kids for a drive round the Island. Perhaps that was a poor choice of words, because unlike Barra that just has one road that goes all the way round the coast. Islay has a lot of roads that go somewhere then come to a dead end, there are a couple of roads that link up and there is a single track that goes down the centre of the island and is an alternative route to Bridegend and beyond from Port Ellen. Bridgend is at the bottom of Loch Indaal and would have made an ideal place for major settlement, as it is the hub of the nearest thing to being a crossroads. It in truth it is not much more than a hamlet even if it could boast a pub, a hotel a store and a church. It could not match the likes of Port Charlotte, Ballygrant let alone Port Ellen or Bowmore in inhabitants, yet prior to the Highland Clearances the most populous place on Islay was the Headland called the Oa; It's soil is the most fertile on the island and has been given over to food production since those times. The centre of the Island is too hilly or too boggy to support much beyond sheep and apart from Port Askaig there are few settlements along the Sound of Islay on the eastern side of the island.

Between Port Ellen and Bowmore on the coast is a beautiful stretch of sand about six miles long called the Strand. The only things to mar its seemingly unspoilt beauty are the Airport to the south, a hotel and golf course. There are traces of peat workings all over the island and I expect that even today the peat will still be used as fuel. Peat on the island is perhaps more famous for giving Islay Malt Whisky's their distinctive flavour. Each distillery produces a whisky that to the connoisseur is as different as chalk is to cheese but for my unrefined taste buds they all taste the same, that still did not stop me from visiting all of the distilleries still producing whisky during my time on Islay and I dare say that given time, the will and a few more visits, I too would have become a connoisseur.

The Lighthouse is on the island of Orsay which is almost a stone's throw from Port Wemyss and close neighbour Portnahaven, now I'm sure that I will upset one settlement or the other by placing one first and not the other. The measure that I am using it its proximity to the island that is all. It used to be that the Church built between the two to serve the two, had two entrances; I had not been inside the church but could imagine one side being for the Port Wemyss lot and the other for Portnahaven. It takes all sorts to make a world.

Talking of churches, our local church in Bowmore was unusual in that it was round; it was built that way, so they say, so that the Devil would not find a corner in which to hide. Whatever the reason it was a nice place in which to seek guidance and worship or just to find solace in the atmosphere that encompassed you once you had passed into its interior.

Islay had its power from the mainland running through cables that went under the stretch of water that separated it, but when there was a failure on the mainland Bowmore had huge standby generators, I'm not sure if they fed the whole island but they certainly fed Bowmore and surrounding villages and hamlets. There was a jetty across Loch Indaal where the fuel for the island was pumped ashore.

Most of the mail was flown in from Glasgow twice daily and that gave Islay the PA postcode, the PA being the sorting office in Paisley, most of the Hebrides had postcodes PA if their mail was sorted in Paisley. Larger items would come by carrier on the ferry. There was a regular bus service that went between Bowmore and Port Ellen and Bowmore and Port Charlotte and Port Askaig. The Post bus serviced hamlets.

There was no helicopter relief for the Rhinns and even though the stretch of water between Orsay and Port Wemyss had a fast current and best negotiated at the turn of the tide, it was generally navigable by a small boat with an outboard. Port Wemyss is a bit of a misnomer because it does not have a port, or anywhere where a boat could safely tie up. Portnahaven has a little beach where boats can be pulled up out of harm's way. As the Rhinns was going to be my last station, I was not overly bothered about ruining my uniform and besides I did have a spare, the one I wore on relief days was now over ten years old and as each year passed it became tighter especially round the shoulders and under the armpits.

I am glad I was the only one going out, as there seemed little room for me my gear and the provisions as well as the boatman. The hardest part was getting the boat into the water, it was a good job that there were a couple of locals around to give us a hand.

As the boat neared the water I was told to clamber in or get a soaking, I never needed telling twice on that account; I had worn the customary inflatable life jacket but would far prefer to stay dry if at all possible. The boat was stable enough and we made the crossing without incident, as did all of the rest of my reliefs.

Orsay is about the same size as Sule Skerry and its features fairly similar, it too had gullies and outcrops of bare rock, but Orsay rose to a higher point where Sule Skerry seemed flatter. Orsay did not have the puffins but had a good-sized colony of fulmars and kittiwakes. Atlantic gray seals would bask on some of the out crops but scurry into the sea on the first signs of your approach. I saw an otter dashing among the rocks near Port Wemyss but beyond a shadow of a doubt the most spectacular sighting I ever made during my service was to see a mixed school of dolphins and pilot whales swimming passed the island, numbering at least thirty. It's hard to count dolphins when all you can see is their dorsal fins so I could have underestimated or over estimated the number. Whatever it was a sight that I thought remarkable.

The Station was not unusual in any way, bearing in mind though that each Lighthouse has its own daylight profile and the Rhinns was a White tower 29 meters high, I still find it hard trying to work in meters but that is how it will appear on marine charts and how the youngsters will read and measure things. It still sounded an emitter like the one at Holburn Head during fog, now if you ask me I could never quite work that one out, the justification for discontinuing the fog signal was because it was deemed that the fog signals were non directional, meaning you could hear it yet not know the exact direction. What's sauce for the goose must surely be sauce for the gander, so why the exceptions? Money that is the exception, it cost nothing to sound a horn or emitter when you have to have a fixed amount of people at the Station anyway. That truth we could all work out for ourselves without being given any illogical line, after all the same conditions applied at their inception as well as their demise, and surely intelligence should tell you that if the sound is getting louder you are likely to be approaching, the simple answer would be to turn away until the sound diminishes. Anyway I could spend the rest of my life chasing my tail over that one, what's done is done, it was technology that finally killed off the service and that cannot be denied.

Just like the Mull, the Rhinns had an array of halogen lights and once they got round to putting new standby generators in and the new monitoring system then the Rhinns too would be automated. I knew, however that I would be there till March 1992 and she would not be automated till 1998.

Just like at Ruvaal, John insisted on staying in the kitchen, so I found myself with plenty of free time, I tried the CB but there were few people to talk to so I sold my rig the next time I was on the mainland. Most of my free time was spent reading or studying for the Open University. In the end I had to make a hard choice about that, and I had to drop out. I had to attend summer course at Sterling University and of course that would coincide with the School holidays, the children had to make enough sacrifices of their own without me adding to them and besides Mags missed me while I was out at the Rock without me being away for an entire month at Sterling.

There was the usual cleaning and painting at the Station but apart from that it was very much routine.

During the warm lazy afternoons, I used to lie on the grass by the gully and watch the fulmars as they surfed on the wind, up the gully to soar just above my head as they passed, and repeat the process over and over. All the while in the background was the sound of the surf as it broke on the rocks below lulling you into sleep that was hard to resist; the wind that provided the birds with a cushion of air was just a gentle breeze at ground level that kept you cool on even the warmest days but would not protect you from the sun's harmful rays, I always kept my body covered and my face with my hat when I dozed in the afternoons.

I remember being entertained for hours one night with one of nature's other phenomena, Aurora Borealis or Northern Lights. The natural light display looks so mystical yet it shows how fragile our earth is and how exposed we are to the harmful side to our life giving sun, if it were not for the protective barrier of our atmosphere then our planet would be as desolate of life as our moon. The sunspot activity that made possible all of the long-range transmissions on the CB could destroy them and all of the other electrical equipment if exposed to that kind of radiation on an extreme level.

They say it is rare to see the Northern Lights so far south, but I could imagine lots of folk in Glasgow watching this spectacle the same as I, that is of course if it was not cloudy. Scotland gets most of its bad weather from the West, listen to the forecasts they will often say that the cloud thickens over Ireland, well there are a lot of hills between Islay and Glasgow and as the coast of Ireland can be seen on clear days from the Rhinns, Islay and the rest of the inner Hebrides could be just as bad a harbinger of bad weather.

I am not the kind of person that ticks off days on a calendar, I feel superstitious about it, and I tend to take mental note of the days but try to keep it as far at the back of my mind as possible. Deep down I fear that something might happen to prevent me from ticking off the final day and I will be left in a kind of limbo for all eternity.

Keeping things and feelings hidden is not such a good idea either, because it makes it harder for you to deal with them later on; and combined with circumstance can lead others to make assumptions that will make them come to wrong conclusions.

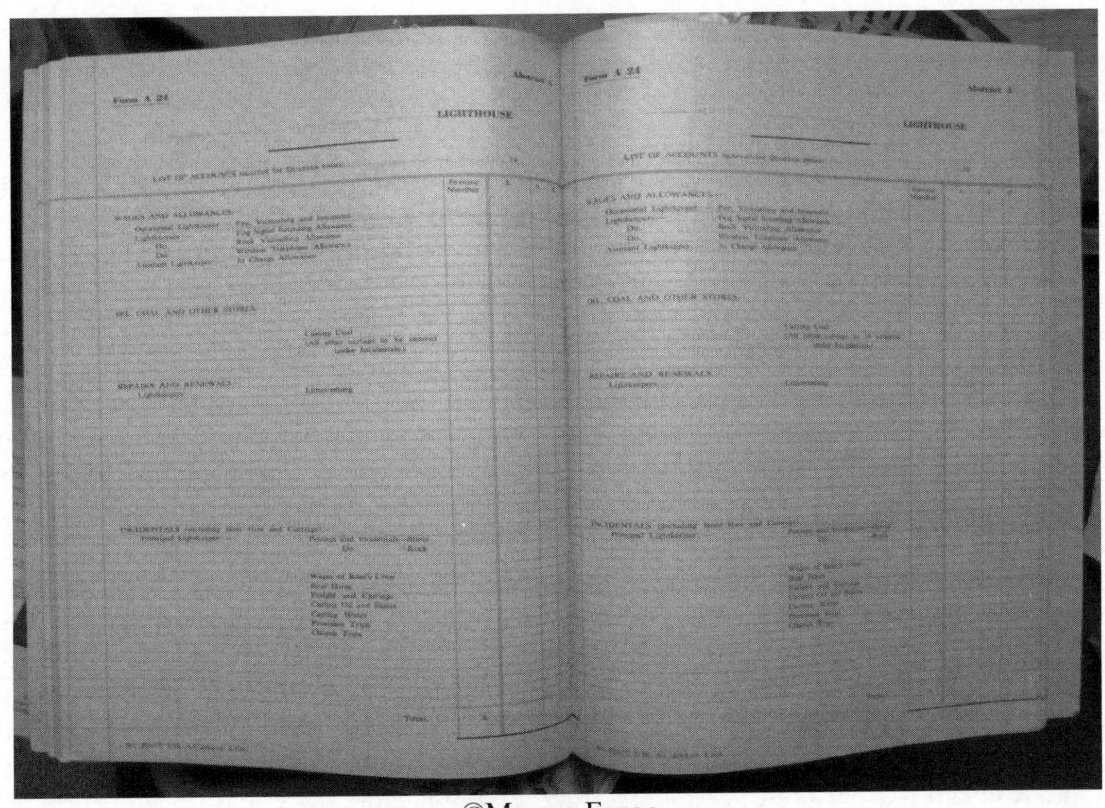
Form A 24

LIGHTHOUSE

WAGES AND ALLOWANCES

OIL, COAL AND OTHER STORES

REPAIRS AND RENEWALS

One of the many Returns sent monthly to 84 George Street

Chapter XXIX
In sickness and in health

I thought of myself as being lucky that I had generally been blessed with good health, yes I got the odd cold now and then and I'd had a few accidents or mishaps that had kept me off work, but nothing major or long term. I had even managed to give up smoking while I was on the Craig; Mags was none too happy at the time because she felt that I had forced her into quitting too. We had talked about giving up for the sake of the children, as both Gavin and Kirsty were asthmatic; talking about it is one thing, committing you to action is another. I reckoned that if I could get passed the first three days I could make it, just like Cortez, I burned my boat by leaving my tobacco ashore. But unlike Cortez, I had the knowledge that should I fail, Alec would be out in a week and if I was really that desperate I could scrounge a few cigs from Norman.

Twenty years later, "do I feel better for it?" I hear you ask. In my mind yes I do feel better for it, but there is another old saying in Scotland…What's before you will not go by you…And that seems especially so when it comes to some illnesses and conditions; some seeds are sown the minute you are conceived and are beyond your ken only to materialise in later life. There is still much to learn about things we can inherit or be predisposed to, just because we carry a certain gene; and yet even among siblings who possibly carry the same gene there can be differences. For example my brothers Bryan and Allan, my sister Susan and I all have angina or heart problems, all caused by a narrowing or furring up of the arteries. Susan has a high cholesterol level, Allan not quite so and I have a level below 5, yet who is the one that has to have a balloon put into their artery to open up the blockage and stents to keep it open. In my mind I say things could have been a lot worse if I had not given up smoking and would have made worse my predisposition to have heart problems. My father and mother died of a myocardial infarction, better known as a heart attack and out of six children three have ischemic atheroma or narrowing of the arteries so maybe once they have mapped the entire G Nome they might find some answers.

Now it may have occurred to you…What the hell does that have to do with Rhinns of Islay…well, directly nothing; but it does lead me gently into what happens next and in the process help me understand the reasoning behind a judgment a certain doctor made.

I had suffered a bit with heartburn and acid reflux when I was out at the Craig but it was nothing that a few pills and a change in diet couldn't make tolerable; notice I avoided the word cure, I did so for a good reason. It was a suspected peptic ulcer, but since it responded to treatment there was no cause to send me to have a barium meal X-ray. At the time they thought ulcers were caused by poor diet and worrying about things was also a contributory factor, perhaps I had been concerned over my future after redundancy and those worries brought on a worsening of the condition. Up until the Rhinns I did not need to have any time ashore because of it and I was determined that was how it should remain. First of all Dr Hardy had to send me to hospital just to see if there was indeed an ulcer causing the problem. Islay does have a small cottage hospital but any specialist X-ray has to be done on the mainland and that involves a flight to Glasgow airport and a short bus trip to Paisley's Alexandra Hospital. Yes I had a small yet significant peptic ulcer. Now Dr Hardy had to find the right treatment, first of all I was prescribed Zantac the same as on previous occasions but that no longer worked, then I went onto Tagamet; that seemed to work for a while, eventually I responded well to Omeprazol; and I have been on it ever since. I was glad that the Tagamet did not work because while I was out at the Rhinns, I got a weird phone call from Mags. She had heard on the news that a man had to have a mastectomy because the Tagamet had given him man boobs and he had developed cancer in one of them.

I'm not on Tagamet but today I do have a fine pair of man boobs and it would be just my luck to go and get cancer in one or both of them.

Discovery of a bacteria that lives in the stomach has done much to advance knowledge about peptic ulcers and radical treatment with antibiotics has rid some patients of the bacteria so that they are less inclined to get ulcers, unfortunately I fall into the category of the tried it and failed; now I am intolerant to penicillin; all this was in its infancy at the time I saw Dr Hardy and as I responded well to treatment there was no need to send me away for tests they would all come after I left the service. So far I still had not had any time off work that was about to change through a silly event that had long lasting consequences.

It was during the summer holidays and we were going to take the girls to Patna and Coylton and spend a few days with Jessie before coming home with Gavin.

There was a sailing from Port Ellen to Kennacraig at 5.30 am and that would mean we could be in Patna before 1pm. The usual ferry had been sent away to Oban to cover the Mull route while the Isle of Mull was being serviced, so we had the old Pioneer. With not so many people travelling at that time of the morning, it would be a pleasant enough trip and we would be able to stretch out and make up for any lost sleep on the 2-hour crossing.

The problem started when we were driving off the ferry. Because of the nature of the tide and the ships ramp the feet at the bottom of the ramp did not quite sit flush on the landing; the feet stuck up just like a knuckle and unfortunately at the same height as my exhaust…Why me… I kept asking as the exhaust parted in the middle. There we were stuck in the middle of nowhere, pissing down with rain, no chance of getting any help for at least a couple of hours what more could go wrong?

I somehow had to connect the exhaust together even if it was only to get us to Lochgilphead I would be happy. The ground was soaking but I had to get under the car, the only bit of good fortune was that if I could get them together one half would slip into the other like a sleeve. Try as a might without proper tools it was going to be a problem whatever method I used. In the end it was down to brute strength and in the process my right thumb got bent back so far that I thought that I had dislocated it. It was pretty darn sore and I thought that the least I had done was sprained it. It worked in as much as the exhaust was now connected and we could drive to Lochgilphead.

There was not much they could do for us there except spot weld the two parts together, that would get us to Ayr and we could get a new exhaust fitted.

Well that was the car sorted but as for my thumb; initially it felt like a sprain but as the days went by it was a sprain that just seemed not to get better. Soon it was time for us to return to Islay and it would be less than a week until I went out to the Rhinns.

For peace of mind more than anything else I needed to find out what was wrong with my thumb, apart from the odd spasm of pain I had little or no grip with it, if I had to do any serious lifting with it; then it would pose a major problem. There was very little going on at the Rhinns that would require me to do any heavy lifting, so I had every intention of continuing until it or I became a liability through it. However, once Dr Hardy knew that I was due to go out to the Rhinns the following week he got this bee in his bonnet that I was swinging the lead and didn't want to go out; he didn't say so, but his judgment had been clouded by those thoughts and I was given Voltarol cream and some unbranded paracetemol as painkillers. I could not gainsay him when he suggested that I could always come ashore if it got any worse but over the course of about a year that was to be my treatment apart from a couple of times when I was given cortisone injections in my thumb to help the joint. So far without knowing the underlying cause of the problem and without having to put the thumb to any real test I was able to go out to the Rhinns without having to come ashore.

Coincidently, it was while I was at Jessie's that I hurt my thumb for a second time and the pain returned with a vengeance. I had little choice but to wait till I got back to Islay the following day, even if it meant me finding a new way to grip the steering wheel and grimacing every time I banged it. This time it was not Dr Hardy that I saw but a Locum, she was immediately alarmed to read from my records that this had been going on for about a year and she asked me if I had ever seen a specialist; that was about to be rectified the instant I came back with a no, she also insisted that I do not return to normal work until I had seen the specialist. She suspected that I could have torn ligaments in my thumb because there was no indication of chipped debris from an X-ray of the joint. I had to tell her without intending any reverse pun that there was no such thing as light duties in the service, you had to be medically fit for work or be signed off. It was even harder for me to plea that as there had been little for me to do out at the Rhinns and that so far I had managed for a year surely I could manage until a true diagnosis had been confirmed. She reluctantly agreed but warned me of the danger of making the injury worse.

Within six weeks I had my appointment with the specialist at the Western General in Glasgow; it would be during my time ashore so I needed no time off work. The health board anticipated that I would not be able to do the trip in one day so had booked me into one of the airport hotels ready for me to catch the first flight back in the morning. He examined my thumb with just the touch of his fingers and announced with regret that no amount of physiotherapy, Voltarol or cortisone would help with this type of injury and all I could do was laugh and then the glisten of a tear would appear. The relief that I was not creating the injury in my mind was almost overwhelming and it was hard to take in what he was about to reveal and his prognosis; I had indeed torn all the ligaments off my thumb so technically there was little to prevent my thumb from dislocating risking all kinds of complications thereafter. What I needed was surgery to reconstruct the ligaments; now came the all important bit, because I had waited so long before getting diagnosed there would be no guarantee that the surgery would be successful, they normally operate within six weeks of the injury, consequently as the chances of a successful operation were severely reduced I would need to wait on a list to have the operation. I could see the logic in that, after all I had managed so far and waiting a bit longer would not reduce my chances of successful surgery. It's no use of going on about If's and and's what was done was done. He gave me a letter addressed to Dr Hardy and the instructions; that under no circumstances was I to risk injuring the thumb further while I was awaiting an operation, so I had to limit it use to minimal activity.

I never said a word to Dr Hardy until he had read the letter and made his response first; apologies are the first steps toward litigation as they admit some kind of liability; so given that he was prone to assumption anyway it was not surprising that one was not forth coming. All I hoped for was that he did the job for which he was trained and not delve too deeply into the psyche of his patients bearing in mind that graveyards the world over, bear testament to the adage that…Assumption is the Mother of all Cock Ups…

©Mearns Forge

Two entries from the works orders and the payments made to keepers

Chapter XXX
Last Days in the Service

Although maths is not my best subject I can still put two and two together and make four. The Locum had already prepared me that I may not be fit for work until I had the operation, so with that in mind I made sure that I limited myself to the necessities while I was out at the Rhinns, just in case the next time would be my last time; as it turned out my last working days in the service were just before I went to see the specialist. I always imagined my last time at the Rhinns would be something special and go out with the bang that Mearns and Bill had planned for me; they enjoyed a good celebration and Christmas and New Year were never dull when you were out with them. There would always be three keepers on watch during the night in the festive season and we would watch the usual crap leading up to the bells then after wishing each other the compliments for the New Year, regale in stories of our passed or jokes that we had recently heard.

Bill was in the RAF and served with the Lochaber Mountain Rescue Team, he told us the story of when they had to go up to retrieve a corpse of a man that had been missing for over a week. They took it to a bothy to clean it up before transporting it down the mountain. The only water to be had was from a burn that was about a ¼ of a mile away. Just like Jack, Bill was sent to fetch a pale of water, unlike Jack he came back in one piece. The newest and youngest member of the team was sent to find a suitable place to radio a sit rep back to base. While he was away the others cleaned up the body; on his return the lad was ordered to make tea while the rest prepared the gear to take him down and get themselves cleaned up. Tea was made and they were all supping happily until it dawned on Bill the pale in the corner was now less than half full and the tea had been prepared pretty quickly. He says to the lad, "Where did you get the water from?" he already guessed the answer but needed confirmation before retching up the contents of his stomach. The only one not to baulk was the Team Leader a Flight Sergeant who happily continued drinking his tea. Bill says to him, "Flight, don't you know where he got the water from?" "Yes", he replies, "but I like my tea with a bit of body in it." Those stories went on all night until we completely forgot who was on watch, anyway it did not matter the light was lit and turning and the sky was clear and the neighbours would not be disturbed in their revelry.

Without ceremony my last days on the Rhinns passed with not even a whimper. I have used the term rock or Rock Station so often that you may be wondering why I have not referred to the Rhinns that way, it's simply did not have the same kind of isolation as the other Stations. Orsay was so close to habitation that you could see people going about their daily business, the only folk you could not see were your own folk. Talking about folk going about their business there seemed to be a lot of activity going on just to the south of Portnahaven; we could see right down this gully and at the far end of it you could just make out a kind of barrier. The work had been going on for nearly a year and then we heard on the news that Belfast University were trying to harness the energy of the surge of the tide to turn it into electricity and had succeeded, I'm not sure if they were the first to generate wave power electricity and connect it to the grid but they must surely have been at the forefront of the technology.

I would be shore bound for the next five months, and then in November we would make our last move back to the mainland. I did not want to go back to Bellsbank and getting a house in Dalmellington was like finding hens teeth, I persuaded Mags that Patna was going to be the best choice and we had our name down for both a council house and a Scottish Homes house. Scottish Homes had housing developments that augmented the council ones especially in the old mining villages in Ayrshire, and it was with them that we were offered a house in Patna that was suitable for our needs. Karen said she detested Islay and was so unhappy that we sent her to live the last six months of my service with Jessie; funny thing was that amongst the family she was the only one ever to return to Islay and that was because her boyfriend's mother came from Islay and they went to visit his relatives.

The Board was sending Keepers near to redundancy or voluntary retirement on vocational training to help them find work and prepare them for life outside the service. It was my turn to go and the course was at the Nautical College in Edinburgh. We had rooms booked in a local hotel for the four days of the course. The board could have saved themselves a bit of money had they waited until we had flitted to Patna, as it was they had to pay for a return flight to Islay. I met a few colleagues and Ross McGilchrist was among them, apart from Edinburgh the last time I had seen Ross was when he appeared on television with three other Keepers on a quiz show called Busman's Holiday. Ross was not too concerned about redundancy as he and his wife had made plans to secure their future. He was lucky, until I had the operation on my thumb I was in a state of limbo so could not lay plans for tomorrow let alone the future.

We had visitors to Islay during the summer, first of all Jessie flew over with Maggie for a week and the weather was great for them, we even took a trip over to Jura and had a picnic in this beautiful picturesque little cove half way round the island. We also paid a visit to the distillery at Coila and saw the ferry pass up the sound on its way to Oban. Then we had a visit from Jennifer and Willy who had been married for about a year now, it seemed like only yesterday that I was carrying on with her while waiting for

Mags to come home from her work; and how she used to say that I made her dizzy, well she has somebody else to make her dizzy now.

I will say, the good thing about Islay is that it is a great place for a holiday, but deep down both Mags and I are semi townies, we like to be not too far away from shops and we would sooner be spoilt for choice rather than have the limited choice that island life seemed to offer; yet we also like village or small community life and could not stand the hustle bustle of city life. Patna would be just fine for us; well I had made the decision so it better just be.

We had an invitation to Pamela's wedding that we gladly accepted; now Mags was presented with two problems, what to wear and what to get for a present. All I got for a few weeks was, " See you…and living on this God Forsaken Island, How am I going to find something to wear?" it would not have mattered one iota had we been living in the middle of Glasgow, I would still have got it in the neck. Just as with everything else we needed while we were living on Islay Mags would just have to look in her catalogue for the solution.

I was not the only one who had to have time ashore, John Bambrick never mentioned it personally to me but I believe he may have had bowel cancer and must have suffered immensely while he was out at the Rhinns and right up until his death. I don't like reading obituaries that is why I am glad that I don't get the Lighthouse Journal, obituaries remind me too much of my own mortality and I would much prefer to think of people in life and as they were.

While we were in Girvan we got news that Ken Clark had passed away; he had moved from Glenruther Lodge when his mother died and went to live in a caravan down in Grantham, it was his cousin who phoned us to give us the news.

Not long after I left the service we heard that John Lamont had passed away too; that seems to be a thing among Lightkeepers; just a lucky few of us live much beyond retirement age. Call me semi mental or even sentimental but I would rather see their light still shining like the Lighthouses they once manned, rather than like old soldiers who never die.

Well my uniform stayed in the Wardrobe until it no longer fitted, I kept the hat and it is still in the loft; and somewhere up there are the buttons I removed from the jacket but what will remain with me for the rest of my days will be my memories of the service.

God Bless all of my fellow Keepers

The End

Forthcoming projects by the author

A collection of lighthouse poems;
here is one from his collection.

Artificers being landed on the automated lighthouse of the Flannan Isles

The rail line last used in 1971 to take the stores from the top landing to the lighthouse

Eilean Mor
The Flannan Isles

Three leagues west of the Lewis Isle
A cluster of Skerries lay in defiance of a westerly sea
Who's raging torrents and stormy swells
Make battering rams against their bastion walls
Thundering as they crash around and over
Cascades topping even mighty Mor
Where I stand in deference to watch in awe

My beam of light picks out
A myriad of diamonds in surf born spray
And spume thrown up from boiling sea
White horses play in a distant field
While swells and troughs roll incessantly
Till they crash against my neighbours pile
Lost to vision as spray fills the air with Neptune's bile

I watch these scenes as the years pass by
A lone audience to an age old play
And no one, net even sons of earth to keep me company
Yet I was built as an aid to men
And tended by them to keep me so
Twice yearly they come and visit me
Barely remembering their coming or seeing them go

Liken to a memory and Echo of the past
While I was but in my infancy
My paintwork was hardly dry and my purpose full of discovery
My keepers too were new to post
Though fashioned by work at other lights
I watch them labour to keep me clean
I see their craft put to the test

The seasons changed from autumn to winter
When Atlantic storms conspired to gather
One after t'other in rages batter
Testing more than mortals in wind and weather
My structure soundly built thus stands
Though battle scarred and weary

My whiteness blasted clean to stone

So it was on an eve like this, just five score years ago
That an angry Neptune came knocking at the door
Seeking vengeance on the souls' of men
Just to even some lost and forgotten score
Sending forth his heralds upon the wind
With deepening swells in hostility
So cruel and callous, bereft of all nobility

The sounds of banging but barely heard
Above the noise of wind and surf
Alerting crew to arrest the damage
A battering door in jam a slamming
And metal to metal of storm lockers hammering
Though mindless of the perils awaiting
Two went to see what they could do

Braving wind and forced to cowl
Shielding faces with gritting scowl
They fought their way to topmost landing
Where locker stood bereft of door
Silenced now with contents scattered
Ropes once coiled, writhed now like snakes
Towards the place where they were standing

They set about the tasks before them
Salvaging as best they could
Securing all in such a fashion
That might protect from further harm
Just as their labours all but done there came a mighty roar
A sound that left them both bewildered
A sound they had never heard before

Suddenly a wall of water had crashed upon the windward side
Sweeping up and over washing clean away
Everything not fixed or tied
My vision all but obscured as the air was filled with blinding spray
What happened to the other man? I simply do not know
I heard him laying a table preparing for their tea
He was safely in the kitchen and I did not see him go

Perhaps it was the sound that drew him out
Seeing the wall of water it left him in no doubt
To his fellows working a word of warning he must shout
But the wall of water hit him
And he was first to meet his doom
So I was left alone in darkness
With no light to beat the gloom

For a human lifetime they had cared for all my needs
Seeing all the things that I saw and on what the spirit feeds
We watched Aurora Borealis
As the lights played through the heavens
With its ghostly hues and rippling waves
To match the calm and peace serene
Of the earth bound ocean scene beneath

We have shared the better days, when Sol has worked her spell
On balmy days her warming glow
Dried their labours as they dressed me all in white
But I am a bride without a groom, and none to share my plight
So I remain a lonely spinster
With my dreams of days to come when man will once more share with me
What natures entertainment can provide